Advances in Intelligent Systems and Computing

Volume 478

Series editor

Janusz Kacprzyk, Polish Academy of Sciences, Warsaw, Poland
e-mail: kacprzyk@ibspan.waw.pl

About this Series

The series "Advances in Intelligent Systems and Computing" contains publications on theory, applications, and design methods of Intelligent Systems and Intelligent Computing. Virtually all disciplines such as engineering, natural sciences, computer and information science, ICT, economics, business, e-commerce, environment, healthcare, life science are covered. The list of topics spans all the areas of modern intelligent systems and computing.

The publications within "Advances in Intelligent Systems and Computing" are primarily textbooks and proceedings of important conferences, symposia and congresses. They cover significant recent developments in the field, both of a foundational and applicable character. An important characteristic feature of the series is the short publication time and world-wide distribution. This permits a rapid and broad dissemination of research results.

More information about this series at http://www.springer.com/series/11156

Mauro Caporuscio · Fernando De la Prieta
Tania Di Mascio · Rosella Gennari
Javier Gutiérrez Rodríguez
Ricardo Azambuja-Silveira
Pierpaolo Vittorini
Editors

Methodologies and Intelligent Systems for Technology Enhanced Learning

6th International Conference

 Springer

Editors
Mauro Caporuscio
Department of Computer Science
Linnaeus University
Växjö
Sweden

Fernando De la Prieta
Departamento de Informática y Automática
Universidad de Salamanca
Salamanca
Spain

Tania Di Mascio
Dipartimento di Ingegneria e Scienze
    dell'Informazione e Matematica
Università degli Studi dell'Aquila
L'Aquila
Italy

Rosella Gennari
Computer Science Faculty
Free University of Bozen-Bolzano
Bolzano
Italy

Javier Gutiérrez Rodríguez
ETS Ingeniería Informática
University of Sevilla
Sevilla
Spain

Ricardo Azambuja-Silveira
Department of Computer Science
    and statistics
Federal University of Santa Catarina
Campus Universitário
    Cx.P. 476, 88040-900
Florianópolis S.C. (Brazil)

Pierpaolo Vittorini
Medicina clinica, Sanità pubblica,
    Scienze della vita e dell'ambiente
Università degli Studi dell'Aquila
L'Aquila fraz. Coppito
Italy

ISSN 2194-5357   ISSN 2194-5365 (electronic)
Advances in Intelligent Systems and Computing
ISBN 978-3-319-40164-5   ISBN 978-3-319-40165-2 (eBook)
DOI 10.1007/978-3-319-40165-2

Library of Congress Control Number: 2016940892

Printed on acid-free paper

This Springer imprint is published by Springer Nature
The registered company is Springer International Publishing AG Switzerland

Preface

Education is the cornerstone of any society and it serves as one of the foundations for many of its social values and characteristics. Knowledge societies offer significant opportunities for novel ICT tools and applications, especially in the fields of education and learning. In such a context, the role of intelligent systems, rooted in artificial intelligence, has become increasingly relevant to the field of Technology Enhanced Learning (TEL). New intelligent solutions can be stand-alone or interconnected to others and can target not only cognitive processes but can also take care of motivational factors. Examples of such solutions are gamified systems, as well as systems based on learning methodologies that stress the role of motivation.

The 6th edition of the mis4TEL conference expanded the topics of the evidence-based TEL workshops series in order to provide an open forum for discussing intelligent systems for TEL, their roots in novel learning theories, empirical methodologies for their design or evaluation, stand-alone solutions or web-based ones. It brought together researchers and developers from the education field and the academic world to report on the latest scientific research, technical advances and methodologies.

This volume presents all papers that were accepted at mis4TEL 2016. All underwent a peer-review selection: each paper was assessed by at least two different reviewers, from an international panel composed of about 50 members of 10 countries. The program of mis4TEL counted 18 contributions from diverse countries, such as Australia, Belgium, Brazil, Italy, Mexico, Portugal, Spain, United Arab Emirates, United Kingdom and the United Stated of America. The quality of papers was on average good, with an acceptance rate of approximately 70 %. Moreover, this volume includes the invited paper of the plenary speaker of the conference, Marcelo Milrad, of the Linnaeus University (Sweden).

Last but not least, we would like to thank all the contributing authors, reviewers and sponsors (IBM, Indra, Fidetia and IEEE SMC Spain), as well as the members of the Program Committee, of the Organising Committee for their hard and highly valuable work. The work of all such people crucially contributed to the success of mis4TEL'16.

June 2016   Mauro Caporuscio
                                                              Fernando De la Prieta
                                                                  Tania Di Mascio
                                                                  Rosella Gennari
                                                        Javier Gutierrez-Rodriguez
                                                        Ricardo Azambuja-Silveira
                                                              Pierpaolo Vittorini

The original version of this book was revised.
An erratum can be found at 10.1007/978-3-319-40165-2_20

Organisation of mis4TEL 2016

http://www.mis4tel-conference.net/

Program Chairs

Mauro Caporuscio	Linnaeus University, Sweden
Fernando De la Prieta	University of Salamanca, Spain
Tania Di Mascio	University of L'Aquila, Italy
Rosella Gennari	Free University of Bozen-Bolzano, Italy
Javier Jesús Gutierrez Rodríguez	University of Sevilla, Spain
Ricardo Azambuja Silveira	Federal University of Santa Catarina, Brazil
Pierpaolo Vittorini	University of L'Aquila, Italy

Program Committee

Silvana Aciar	National University of San Juan, Argentina
Ana Almeida	Institute Pedro Nunes, Portugal
Peter Bednar	University of Portsmouth, UK
Margherita Brondino	University of Verona, Italy
Edgardo Bucciarelli	University of Chieti-Pescara, Italy
Davide Carneiro	Universidade do Minho, Portugal
Pablo Chamoso	University of Salamanca, Spain
Vicenza Cofini	University of L'aquila, Italy
Juan Cruz-Benito	University of Salamanca, Spain
Giovanni De Gasperis	University of L'Aquila, Italy
Vincenzo Del Fatto	Free University of Bozen-Bolzano, Italy

Local Organising Committee

María José Escalona Cuaresma (Chair)	University of Sevilla, Spain
Carlos Arevalo Maldonado	University of Sevilla, Spain
Gustavo Aragon Serrano	University of Sevilla, Spain
Irene Barba	University of Sevilla, Spain
Miguel Ángel Barcelona Liédana	Technological Institute of Aragon, Spain
Juan Manuel Cordero Valle	University of Sevilla, Spain
Francisco José Domínguez Mayo	University of Sevilla, Spain
Juan Pablo Domínguez Mayo	University of Sevilla, Spain
Manuel Domínguez Muñoz	University of Sevilla, Spain
José Fernández Engo	University of Sevilla, Spain
Laura García Borgoñón	Technological Institute of Aragon, Spain
Julian Alberto García García	University of Sevilla, Spain
Javier García-Consuegra Angulo	University of Sevilla, Spain
José González Enríquez	University of Sevilla, Spain
Tatiana Guardia Bueno	University of Sevilla, Spain
Andrés Jiménez Ramírez	University of Sevilla, Spain
Javier Jesús Gutierrez Rodriguez	University of Sevilla, Spain
Manuel Mejías Risoto	University of Sevilla, Spain
Laura Polinario	University of Sevilla, Spain
José Ponce Gonzalez	University of Sevilla, Spain
Francisco José Ramírez López	University of Sevilla, Spain
Isabel Ramos Román	University of Sevilla, Spain
Jorge Sedeño López	University of Sevilla, Spain
Nicolás Sánchez Gómez	University of Sevilla, Spain
Juan Miguel Sánchez Begines	University of Sevilla, Spain
Eva-Maria Schön	University of Sevilla, Spain
Jesús Torres Valderrama	University of Sevilla, Spain
Carmelo Del Valle Sevillano	University of Sevilla, Spain
Antonio Vázquez Carreño	University of Sevilla, Spain
Carlos Torrecilla Salinas	University of Sevilla, Spain
Ainara Aguirre Narros	University of Sevilla, Spain
Diana Borrego	University of Sevilla, Spain
Fernando Enríquez de Salamanca Ros	University of Sevilla, Spain
Juan Antonio Alvarez García	University of Sevilla, Spain
Antonio Tallón	University of Sevilla, Spain

Contents

Part I
Invited Paper

Web Technologies and Mobile Tools to Support Sustainable Seamless Learning

Marcelo Milrad

Abstract In this paper I present and discuss our efforts and experiences with regard to those aspects connected to the design and implementation of seamless learning activities and how to promote those in order to become sustainable. A particular case in the field of mathematics education is presented and discussed. In terms of technologies to support seamless learning we promote the use of web-based, mobile tools and reusable and flexible components, as they allow teachers and students to create, adopt and adapt learning activities that connect the physical and the digital world across different locations and settings.

Keywords Mobile and seamless learning · Web technologies · Mobile tools and systems · Reusable components

1 Introduction

The field of Technology Enhanced Learning (TEL) has been continuously evolving during the last three decades. The introduction of computer-based training, and later on networked-based learning, mainly due to the advent of the World Wide Web, led to the definition of *e-learning*. Advancements in mobile and wireless technologies have also had an impact in educational settings, thus generating a new approach for technology-enhanced learning called mobile learning or *m-Learning* [1]. The rapid development of these latest technologies combined with access to content in a wide variety of settings, allows learners to experience new learning situations beyond the school's walls. This latest view on technology-enhanced learning supported by wireless technologies and ubiquitous computing is referred to Ubiquitous Learning or *u-learning* [2]. While context is an important

M. Milrad(✉)
Department of Media Technology, Linnaeus University, 351 95, Växjö, Sweden
e-mail: marcelo.milrad@lnu.se

© Springer International Publishing Switzerland 2016
M. Caporuscio et al. (eds.), *mis4TEL*,
Advances in Intelligent Systems and Computing 478,
DOI: 10.1007/978-3-319-40165-2_1

aspect of mobile learning, it is the central concept of u-learning, due to two important features of the learning environment, namely *context awareness* and *adaptivity*. The notion of *context awareness* means that the pedagogical flow and content that are provided to the learning environment should be *aware* of the situations in which the learner/s actually is/are. The term *context adaptivity* refers to the idea that different learning contents should be adaptable to the particular setting in which the learners are situated.

This latest view on technology-enhanced learning offers the potential for a new phase in the evolution of technology-enhanced learning, marked by a continuity of the learning experience across different learning situations. Chan and colleagues [3] and Milrad et al., [4] use the term *"seamless learning"* to describe these new situations. Seamless learning implies that students can learn whenever they are curious in a variety of situations, they can easily and quickly s witch from one scenario to another using their personal mobile device as a mediator, and can maintain the continuity of their learning across technologies and settings. These scenarios include learning individually, with another student, a s mall group, or a large online community, with possible involvement of teachers, relatives, experts and members of other supportive communities, face-to-face or in different modes of interaction and at a distance in places such as classroom, outdoors, parks and museums. Recent studies on seamless learning have been extending from teacher-facilitated classroom or outdoor learning into nurturing autonomous learners [5]. Indeed, the ultimate motivation for learning scientists to promote seamless learning is to foster the habits of mind and abilities that support 21st century skills among students. Thus the aim is to design and enact not just episodic activities but programs to gradually transform learners into more self-directed individuals being able to carry out learning tasks not just anytime and anywhere, but perpetually and across contexts without external facilitations. Mediated by technology, a seamless learner should be able to explore, identify and seize boundless latent opportunities that her daily living spaces may offer, rather than always being inhibited by externally defined learning goals and resources [6].

More research is certainly needed to improve our knowledge to better guide the pedagogical design of seamless learning activities, as well as its technological support. It is necessary to further investigate how students interact with learning contents, peers, teachers and parents through a variety of technologies and contexts and how these innovations can be integrated into everyday education practices in order to become sustainable. The main purpose of the ideas presented in this paper is to illustrate how web and mobile technologies can shape and create innovative ways to share and construct information and knowledge in both formal and informal educational settings to promote seamless learning. One concrete example of our projects is presented and discussed to illustrate the ideas above. Having these points in mind, I discuss in the final section those aspects related to design approaches and technological solutions that can support sustainable seamless learning.

2   Motivation and Challenges

The idea that new Information and Communication Technologies (ICT) will transform learning practices has not yet been fully realized, especially with regard to ICT
to support learning and collaboration across contexts. The task of designing effective
computer support along with appropriate pedagogy and social practices is more
complex than imagined [7]. The design of systems and technological tools to support collaboration and communication in seamless learning environments is a difficult process, not only because the learners may be separated by time and space, but
also because they may be also not sharing the same learning physical context [6].
Establishing common ground and mutual understanding; two important ingredients
for collaborative learning, becomes a challenge. One of the major challenges for
educational technologists and researchers is to find useful ways to design, to implement and to evaluate web and mobile technologies and innovative pedagogical ideas
in a wide variety of educational settings. Some of the current design challenges
faced by educational technologists can be enumerated as follows:

- *How to design seamless learning activities that support innovative educational
  practices and can sustain over time?*
- *How to design seamless learning activities that integrate learning in informal
  and formal settings?*

Another central challenge still lies in the integration not only between software
components in distributed environments, but also in the combination of software
with new hardware and peripherals (e.g., sensors), as well as the support for content
delivery on diverse types of devices used across different educational contexts.
Some of the current technological challenges in this area can be formulated as
follows:

- *What features and capabilities should mobile tools and web-based systems
  provide to support different collaborative learning activities?*
- *How to create adaptable computational mechanisms that would enable personalization and reusability of the learning content for different users across different platforms and tools?*

During the last six years, our research group at Linnaeus University (CeLeKT –
Center for Learning and Knowledge Technologies) has been exploring new design
approaches and innovative uses of social media, web and mobile technologies in a
variety of collaborative and inquiry-based learning settings [8; 9; 10; 11]. These
research efforts are not simply characterized by the provision of novel uses of rich
digital media combined with web and systems and tools, but also by the exploration of new and varied learning activities that become available while applying
innovative approaches for designing new technological solutions and utilizing
existing ones to support seamless learning. These activities have all been closely
collaborating with schools teachers, students and companies with the goal of not

only implementing these innovative mobile learning activities in the schools but also having in mind how these should be maintained and sustained over time. The main objective of the coming section is to exemplify one of the cases we have been working on and the challenges and problems associate with the introduction of innovative learning activities in schools that characterize sustainable seamless learning research and practice.

3   Sustainable Seamless Learning in the Field of Math Education

3.1   The Geometry Mobile (GEM) Project

Geometry Mobile (GEM) is an on-going seamless learning project in the field of mathematics trying to find alternative ways to support the learning of geometry using mobile and positioning technologies [8]. The project brings together a group of researchers from Linnaeus University (LNU) in the fields of media technology and mathematics education working very close with teachers and schools. The activities taking part in the project are related to inquiry-based geometric learning tasks involving transitions between different contexts including outdoors and classroom tasks. The research focus is not only on the appropriation of technologies introduced to the students and teachers but also on how the use of mobile technologies support these transitions, in particular how they support effective communication of mathematical strategies. Moreover, one other objective is how to integrate these learning innovations into everyday educational practices, so they can sustain over time [12]. The project has been motivated by the idea of designing learning activities capable to stimulate students′ enactive mode of action by means of spatial visualization. Guided by design-based research and the notion of seamless learning [3], we have designed and implemented a series of learning activities in mathematics where mobile and web technologies support transitions between outdoor and indoor learning contexts [8]. We have developed a set of mobile applications, which allows a student to measure distances between her own device and mobile devices held by other students, as well as to collect data and record audio annotations. The data collected by these mobile applications is stored in a central repository for using it later in the classroom. A web-based geo-visualization tool and an augmented reality application are used back in the classroom to visualize and reflect upon the activities conducted outdoors in the field.

Since 2009, the research group at LNU has conducted trials and empirical studies with several classrooms from four different elementary schools involving more than 400 elementary students from the south of Sweden. Among the outcomes of the studies, it was observed that by participating in the activity, students are offered opportunities to experience geometrical constructions in full-sized space. Specifically, they are stimulated to make use of their orientation ability, which differs cognitively from the visualization ability, which is more commonly used to solve similar tasks in school. These kind of learning activities are offering the participating students enacted experiences of school geometry that are not commonly offered in school contexts [8]. However, these types of activities are high demanding as they are new for the teachers and the learners who need to adjust or even recreate adequate frameworks for communication and interaction with and through new technological devices. Furthermore, the fact that these activities are outdoor explorations, require the use of mobile technologies, and organizing the activity across time and locations poses didactical as well as technological challenges for the teachers and the school that call for careful considerations regarding the design of innovation in schools [14].

Regarding sustainability aspects of the pedagogical and technological activities that the GEM project brings to the school, we observed the complexity inherent to the initiative. Complexity that is partly due to the multiplicity of voices, interests and relationships involved in such seamless learning activities (i.e. the teachers, the students, students' parents, researchers, designers, programmers, developers, school director and IT-support personnel at the school, municipality representatives). This type of initiatives are thus new for the school; but not only are novel because the school opens up its doors to new technologies -mobile devices, interactive visualizations- it is new in the sense that the school opens up its doors to a new social universe of relationships, roles, responsibilities, competence. From a technological perspective, one major challenge while implementing the different approaches lies in the fact that the field of TEL is lacking a cross-platform mobile solution to meet the requirements of the fragmented mobile market. Analysis components like the one presented in [10] allow to reflect upon a performed learning activity by representing the data that is collected during an activity in a way that users can draw conclusions from it. The central idea behind an analyses component is to enable students to easily perform a visual exploration and analysis of the data they have collected to facilitate the understanding of a particular phenomenon and to communicate findings and therefore widens the learning experience.

Based on what we have learnt and the analyses of our previous research efforts in this field we can claim that web and mobile technologies can support different reflection and discussion processes that allow students to understand concepts in the field of geometry, and teachers have shown good skills related to efficient use of the different software solutions [14]. However, the appropriation and adoption of web and mobile technologies by teachers may not be enough to allow them to design and conduct a seamless learning activity. The different applications should be able to transfer the data that has been collected in groups in the outdoor phase (using mobile devices) to a reflection phase that can be performed by the whole class in the classroom environment (via a desktop computer). Teachers cannot be expected to have the required ICT skills needed for transferring the data between data between different devices and to create tools that can process these data for supporting follow-up discussions. Therefore, their dependency on researchers and software developers becomes a limitation for their future independent use of the technologies. In order to address this problem, the TriGO environment [15] has been conceptualized and implemented and it can be seen as an evolution of GEM. A detailed description of TriGO is provided in the coming section.

3.2 TriGO: The Next Evolution of the GEM Project

TriGO is one of our latest developments and natural evolution resulting from the outcomes of the GEM project. One big challenge that the TriGO environment is addressing relates to the design and the preparation of the learning activity itself. In the context of our efforts a seamless learning activity consists of three main phases, namely *Design, Experimentation and Reflection.* The Experimentation phase starts with an introduction phase in which the teacher introduces the activity and its content to the students before carrying out the fieldwork. In the previous activities of the GEM project, a developer was required in order to work together with a teacher to define the exact locations of the landmarks and put those in a XML-file. Thereafter, those XML-files needed to be copied to the mobile devices that were used during the activity. As a way to facilitate this process and allow teachers to work independently, we have solved this problem by providing an authoring tool that allows users to place a rectangle on an interactive map and then to assign six landmarks to its locations. The technological setup of the authoring tool relies on the reuse and adaptation of the mLearn4web platform [16]. MLearn4web uses the latest web technologies (HTML5, NodeJS, Web sockets) for providing a platform that in an adaptive and adaptable manner allows for the creation of an authoring tool to address the new requirements of the TriGO project. Each activity designed by a teacher is assigned with an ID and a password that needs to be entered in the TriGO mobile application. Once this data is entered into the mobile device allows the corresponding data set to be loaded into the application. By adding the authoring tool, teachers can in an independent way to prepare and implement the activity by themselves without the need of a developer. In this way, it becomes easier to allow the ideas and activities of this project to be disseminated to a big number of schools. Complementary to the authoring tool, a visualization tool was also implemented. Once a particular activity has been performed, it is possible to upload all data sets that are created by the students and visualize them to find a certain point. All this information can then be interactively presented to analyze the strategies that students have selected to find a certain location. These strategies are very important from a learning perspective as they can be used to find a spot first that matches one of the desired distances and then go in a circle toward the second landmark until the desired spot is reached. This entire process, for example, can be visualized by limiting the visualization to a certain group of students and a specific task. The set of applications tools used to support these activities is depicted in Fig 1.

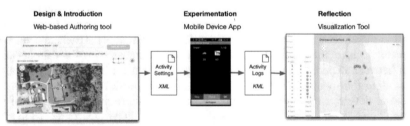

Fig. 1 The authoring tool, the mobile application and the visualization tool components of the Trigo platform

As already indicated the ideas and outcomes of these activities have been extensively evaluated in the last 4 years [14]. However, the introduction of the authoring tool and its impact was not included in those evaluations. In order to explore how teachers experience working with the tool, a workshop with ten teachers has been carried out to validate whether it was possible for the teachers to define the locations where the activity should take place and conduct the activity by themselves. The development of the three blocks of the TriGO application has been performed at different stages and following an iterative development method. Following these activities, we have run assessment studies independently for each one of these three blocks in order to explore their usability and to refine our concepts. The authoring tool created for TriGO was developed guided by the results obtained from a prior study in the use of generic authoring tools. In a previous effort, a generic authoring tool was presented in several workshops to a total of 29 teachers at the Kronoberg region in South East Sweden. By carrying out these studies, we aimed at identifying, first, the user's experiences in using web technologies for designing seamless learning activities and, second, their specific needs for creating those. The results showed that teachers with limited technical skills rapidly learned how to make use of the authoring tool functionalities and did not have difficulties to use the web-technologies to design their own seamless learning activities. The results of all these efforts point out to the fact that the web-based solution in the form of an authoring tool provides a convenient approach for allowing teachers to design their own seamless learning activities. They also enable them to coordinate the preparation for the outdoors phase via the dashboard. Additionally, the web-based solution we are promoting reduces potential problems with the installation of mobile applications. It is important to mention that schoolteachers may not have administration privileges that could be required to install mobile applications. Our group has developed the mobile application used in TRIGO, which it has evolved through the different case studies performed during the last 5 years [15].

In the most recent version of the mobile application, 27 students from grade 4th and 26 students from grade 6th have used the TriGO in two different sessions during 2015. The main rationale of these efforts was to explore usability issues of the application in authentic settings, in which the schools had not access to the research team and the students had not prior training in how to use the mobile application. The first contact that the students had with the TriGO mobile application was ten minutes before the experimentation phase, in which the teacher handled the phones while describing that they were going to participate in a mathematics activity using mobile devices. Based on the recordings and data we collected, the students did not have major difficulties to interact with the mobile application interface, neither to understand the activity flow and how to proceed during the session. Additionally, the seamless interaction and data exchange between the mobile device and the web application seems to reduce the complexity and time required to prepare the mobile devices for the outdoor phase if compare with previous efforts. These different interactions offer the possibility to support a seamless transition between locations, activities and devices [4].

4   Discussion and Concluding Remarks

In this paper I have presented and discussed our experiences and reflections with regard to those aspects that need to be considered in order to design and implement seamless learning activities and to promote those to become sustainable. A particular case in the field of mathematics education has been presented and discussed. In terms of technologies to support seamless learning we promote the use of web-based, mobile tools and reusable and flexible components, as they allow teachers and students to create, adopt and adapt learning activities that connect the physical and the digital world across different locations and settings. Two important components that may differ from similar efforts in this field is the use of a web-based authoring tool to allow teachers to create and design their seamless learning activities, as well as an interactive visualization tool to support multimodal representations and discussions based on the data collected by the students using their mobile devices.

The combination of the authoring tool, the mobile application and the interactive visualization can support seamless transitions across locations and devices and offer the users the possibility to explore learning context in informal settings. The outcomes of our efforts indicate also that the combination of these technologies to support the different seamless learning phases can simplify the challenges that teachers currently face while designing the different activities.

Two perceptible changes in the field of educational research have been recently identified by [17]. The first is a shift in our sense of the spaces and contexts in which education takes place, as different learning activities are becoming more commonly distributed across a variety of contexts. The second change is a wider understanding with regard to the conception of educational praxis, acknowledging the growing importance of design. Addressing these two challenges calls for new integrated design approaches for technology-enhanced learning. One of the major challenges of today's education is no longer about finding the best ways for knowledge delivery, but rather designing, developing and implementing interactive learning experiences and activities for learners to construct knowledge by engaging and inspiring them to learn. We believe that the notions and the concrete example presented in this paper represent a step forward towards achieving such goals. Seamless learning activities of the type illustrated in this chapter, that are highly self-regulated and involve collaboration and communication with peers, can contribute to preparing students for a future which requires them to take initiatives, be creative, take informed decisions, and puts high demand on their social skills. Moreover, the issue related to the sustainability of these type of efforts is quite a complex problem, which is far from being trivial, as there are so many different stakeholders and variables involved. We agree with Pedró [18] that points out to a more systematic approach for understanding technology-based school innovations. He claims that *"There is a need to know more about how governments promote, monitor, evaluate and scale up successful technology-based or supported innovations, paying particular attention to the role played by research, monitoring and evaluation, and the resulting knowledge base, both at national and international level.* [18]". We strongly encourage the international TEL research community to start developing collaborative efforts to promote new ways for thinking and taking concrete actions in these directions.

References

1. Sharples, M., Milrad, M., Arnedillo Sánchez, I., Vavoula G.: Mobile learning: small devices, big issues. In: Balacheff, N., Ludvigsen, S., de Jong, T., Lazonder, A., Barnes, S. (eds.) Technology Enhanced Learning: Principles and Products. Springer, Heidelberg (2009)
2. Rogers, Y., Price, S.: Using ubiquitous computing to extend and enhance learning experiences. In: van Hooftk, M., Swan, K. (eds.) Ubiquitous Computing in Education: Invisible Technology, Visible Impact. Lawrence Erlbaum Associates, Inc. (2006)
3. Chan, T.-W., Milrad, M., et al.: One-to-one technology-enhanced learning: an opportunity for global research collaboration. Research and Practice in Technology Enhanced Learning Journal **1**(1), 3–29 (2006)
4. Milrad, M., Wong, L.-H., Sharples, M., Hwang, G.-J., Looi, C.-K., Ogata, H.: Seamless learning: an international perspective on next generation technology enhanced learning. In: Berge, Z.L., Muilenburg, L.Y. (eds.) Handbook of Mobile Learning. Routledge, New York (2013)
5. Wong, L., Looi, C.: What seams do we remove i n mobile-assisted seamless learning? A critical review of the literature. Computers & Education **57**(4), 2364–2381 (2011)
6. Wong, L.-H., Milrad, M., Specht, M. (eds.): Seamless Learning in the Age of Mobile Connectivity. Springer (2015)
7. Stahl, G.: Contributions to a theoretical framework for CSCL. In: Proceedings of CSCL 2002, pp. 62–71. Lawrence Erlbaum, Hillsdale (2002)
8. Sollervall, H., Otero, N., Milrad, M., Vogel, B., Johansson, D.: Outdoor activities for the learning of mathematics: designing with mobile technologies for transitions across learning contexts. In: Proceedings of the 7th IEEE International Conference on Wireless, Mobile and Ubiquitous Technologies in Education, Takamatsu, Japan (2012)
9. Alvarez, C., Salavati, S., Nussbaum, M., Milrad, M.: Collboard: Fostering new media literacies in the classroom through collaborative problem solving supported by digital pens and interactive whiteboards. Computers & Education **63**, 368–379 (2013)
10. Vogel, B., Kurti, A., Milrad, M., Johansson, E., Müller, M.: Mobile inquiry learning in Sweden: development insights on interoperability, extensibility and sustainability of the LETS GO software system. Educ. Technol. Soc. **4522**, 43–57 (2014)
11. Nordmark, S., Milrad, M.: Influencing everyday teacher practices by applying mobile digital storytelling as a seamless learning approach. In: The Mobile Learning Voyage-From Small Ripples to Massive Open Waters, pp. 256–272. Springer International Publishing (2015)
12. Pargman, T., Milrad, M.: Beyond Innovation in Mobile Learning. In: Mobile Learning: The Next Generation (2015), pp. 154–178. Routledge (2016)
13. Heer, J., Viégas, F.B., Wattenberg, M.: Voyagers and voyeurs: supporting asynchronous collaborative information visualization. In: Proceedings of the SIGCHI Conference on Human Factors in Computing Systems, pp. 1029–1038 (2007)

14. Sollervall, H., Gil de la Iglesia, D.: Designing a didactical situation with mobile and web technologies. In: Proceedings of 9th Congress of European Research in Mathematics Education (CERME9) (2015)
15. Gil de La Iglesia, D., Sollervall, H., Zbick, J., Delgado, Y.R., Sirvent Mazarico, C.: Combining web and mobile technologies to support sustainable activity design in education. In: Proceedings of the Orchestrated Collaborative Classroom Workshop 2015, pp. 1–4 (2015)
16. Zbick, J., Nake, I., Jansen, M., Milrad, M.: mLearn4web: a web-based framework to design and deploy cross-platform mobile applications. In: Proceedings of the 13th International Conference on Mobile and Ubiquitous Multimedia MUM 2014, pp. 252–255. ACM, Melbourne (2014)
17. Goodyear, P.: Emerging methodological challenges. In: Markauskaite, L., Freebody, P., Irwin, I. (eds.) Methodological Choice and Design, vol. 9, Part 4, pp. 253–266. Springer Netherlands (2011)
18. Pedró, F.: The need for a systemic approach to technology-based school innovations. In: Inspired by Technology, Driven by Pedagogy: A Systematic Approach to Technology-Based School Innovations. OECD report, 2010 (2010)

Part II
Methodologies for Fostering Motivation and Engagement

1

An Agent Model of Student's Affect for Adapting Schooling Strategies

Bexy Alfonso, Elena del Val and Juan M. Alberola

Abstract In this paper, we present a student's affective model that considers a temporal and multi-dimensional view of the student. It considers three dimensions (i.e., individual, environmental, and social dimension), which contain static and dynamic features. Based on this model, we define a MAS, that includes emotional agents able to simulate student's and their affective state. This system allows to simulate the effects of changes in lessons over the affective state of students.

Keywords Cognitive · Emotion · Affect · Agents · Modeling

1 Introduction

Affective characteristics (such as emotions) play an essential role in education. They influence students' and teachers' interest, engagement, and achievement, and, in a more general level, the well-being of students. The main question is, how can we adapt education? More specifically, how can teachers use student's affective information in order to improve teacher's schedule and schooling strategies in the teaching sessions, and hence, improving students-related learning factors such as: satisfaction, percentage of comprehension, knowledge speed acquisition, and motivation? Practical considerations on how to modify students' and teachers' emotions can be extracted from the analysis of the results obtained from classrooms observations or from theoretical considerations. Therefore, in this paper, we propose a model that helps teachers to assess what can be done to prevent or reduce negative emotions (i.e., anger, hopelessness, or boredom) in students, and to promote their positive emotions (i.e., hope, pride, and enjoyment) in teaching and learning.

Emotions should be addressed from a multi-dimensional perspective by "addressing the variation of emotions between individuals, activities, and subject domains,

B. Alfonso · E. del Val · J.M. Alberola(✉)
Universitat Politècnica de València, Valencia, Spain
e-mail: {balfonso,edelval,jalberola}@dsic.upv.es

© Springer International Publishing Switzerland 2016
M. Caporuscio et al. (eds.), *mis4TEL*,
Advances in Intelligent Systems and Computing 478,
DOI: 10.1007/978-3-319-40165-2_2

as well as classrooms" [1]. For instance, by analyzing whether the variation of emotional experiences can be explained by differences between academic activities, like attending to a class vs. taking exams, or between different academic subjects, and to what extent. It is also important to determine which of these variables play a critical role on the variation of classrooms' emotional climate. These issues are of fundamental importance for adequately designing educational interventions [1].

In this paper, we propose a computational approach that offers a practical way of evaluating the impact of different schooling strategies on the students affective state. The computational approach is based on an agent model that integrates relevant factors that influence the affective state of a student. The aim of the approach proposed in the paper is to facilitate the estimation of the impact of schooling strategies with the minimum intervention of students.

The paper is organized as follows. Section 2 presents proposals that analyze the influence of emotions in learning process. Section 3 describes the agent model that integrates individual, environmental, and social factors that influence in the student's affective state, and therefore, in his/her learning process. Section 4.1 defines the stages of the process for gathering information in real learning environments and for including it in a computational approach. The computational approach is based on a multi-agent system that facilitates the simulation of the effects of schooling strategies over the students affective state and learning process.

2 Related Work

The influence of affective characteristics on the cognitive processes and behavior of an individual has been systematically demonstrated across several disciplines. Emotions, personality or mood are among the affect-related cognitive concepts that have been more widely addressed [2, 3]. Emotions are known to be the result of an appraisal of certain stimuli that can be events, objects, or other individuals [4, 5, 6]. Personality can be defined as the "dynamic and organized set of characteristics possessed by a person that uniquely influences his or her cognitions, motivations, and behaviors in various situations" [3]. Mood, as emotions, characterizes the affective state of an individual but mood's intensity is lower than emotions' intensity and it represents long-lasting affective states, while emotions have a brief duration [2]. In particular, these affect-related concepts and their influence on cognition and behavior, have been widely investigated for teaching and learning. One of the domains that has most widely studied affect on teaching and learning is e-Learning. For example, in [7] authors propose a model for predicting an agent emotional reaction in a distant environment of learning by using personal characteristics and non-personal data. Also, S. Chaffar *et al.* created a methodology that also starts from the learner's individual information (personality, and motivation) and environmental factors (tutor's intervention type) for predicting the learner's emotional reactions by using a Naïve Bayes classifier [8]. Also, a model for selecting a suitable virtual classmate on the base of student's personality and emotions is proposed in [9].

The Control-Value Theory of Achievement Emotions of R. Pekrun *et al.* [1], is maybe the approach that most closely resembles ours. This theory proposes an appraisal model for ongoing achievement activities, their past and future outcomes where environmental, social and individual factors are appraised. The theory also focuses on the influence of emotions on academic engagement and performance. We focus instead on finding the linkage between the combination of several dimensions (individual, environmental, and social), and a variation on the students emotional state Also, our approach looks for schooling methodologies that better fit student's requirements and current situation, by switching different schooling methodologies and techniques, on the base of an "improvement" of the students' emotional state. It offers a practical way of evaluating the impact of different schooling strategies on the students affective state, and hence in their well being in relation to those strategies. This way the probability of a better learning process and a better students performance grows, as proved by several previous studies and results. By focusing on particular and comprehensive aspects and on a particular design, our approach makes possible the implementation of a computational system able to predict the impact of schooling strategies with the minimum intervention of students.

3   Student's Affective Model

We propose a student's model that considers a set of circumstances that influences in the affective-state of individuals and, therefore, in the effects over a set of learning factors that is interesting to evaluate. The set of circumstances is composed of three dimensions: *individual*, *environmental*, and *social*. In each dimension, we consider relevant features that maintain their value during a long period of time (i.e., course, school year or semester), and features that are dynamic and change their value with a higher frequency (i.e., during a lesson).

The *individual dimension* has two static features: concerns and personality. Concerns represent the individual *achievement goals* of the student in the long term (i.e., learn as much as they can about a subject, personal satisfaction, doing the minimum effort to pass the course, or doing well on exams to obtain a high mark). The relationship between personality and learning is largely documented [10]. We considered the Big Five personality traits [11], in order to characterize the student personality. Combinations of Big Five traits have also been found to predict various educational outcomes. The individual dimension has one dynamic feature: expectations. People tend to behave in such a way that their behavior optimally matches their expectations. The knowledge of student expectations can be also useful for adapting the design of teaching programs or the teaching strategies and methodologies [12]. For instance, if teachers know what their students expect, they may be able to adapt their behavior to their students' underlying expectations, which should have a positive impact on their levels of satisfaction [13].

In *environmental dimension*, the static set of features are those environmental circumstances around the learning context that remain constant during a course or semester. In the dynamic set of features, we consider features that change their value more frequently such as the teaching methodology used by the teacher (i.e., teacher-centered approach or student-centered approach), the position of the lesson in the school year schedule, scheduled events (i.e., exams, holidays, deliverables, etc.), the task performed during the lesson (i.e, the provision of realistic, challenging, or appropriate tasks, etc.).

The *social dimension* is characterized by a set of dynamic features. The classroom climate is one of the factors to consider in the social context and can be classified in: (a) consistently positive and supportive, (b) consistently negative and nonsupportive, and (c) ambiguous. The students-teacher relationship has an effect on student's emotions. For instance, teachers and students can create positive climates for learning alignment between a person's goals and the goals of the classroom [14]. The relation between a student and a peer group is also important and influence in students feelings and emotions such as those related to the achievement success or failure, as well as acceptance or rejection by others [15], or the alignment of peers' academic aspiration [16].

These three dimensions compose the circumstance of the student that has a direct influence over the affective-state of the student. The emotional part of the student has been represented in the model through the affective-state. According to [17], affect seems to be a best term to describe emotions over time. The model includes two ways of establishing the student's emotion: through the PAD model (Pleasure, Arousal, and Dominance), and through an emotion aggregation of emotions that are directly related to students and learning environments. The affective-state of the student influences in *factors in the learning context*. These factors are classified in the model as short-term factors (i.e., students satisfaction, percentage of comprehension, knowledge speed acquisition, and motivation) and long-term factors (i.e., academic results, and satisfaction).

4 Workflow

In this section, we describe a general workflow stages and their most important features where the student's affective model and the proposed computational model are integrated. The stages are divided into two main parts (see Figure 1): *real learning environment* and *simulation*.

Real learning environment includes the set of stages that are required to collect information about students' features that emerge during a lesson. In Stage 1, students fill a survey with information about their individual dimension and the teacher establishes some features of the environmental dimension. Before a lesson starts, students express their affective state using a PAD model in Stage 2, and then, the lesson starts. During Stage 3, the lesson is performed. In Stage 4 (which takes place at the end of the lesson), students provide information about their affective state through a form that includes a set of emotions related to learning. Then, before the next lesson, students can update information about their social dimension or their expectations, which corresponds to Stage 5. The teacher can modify environmental dimension in order to see how changes

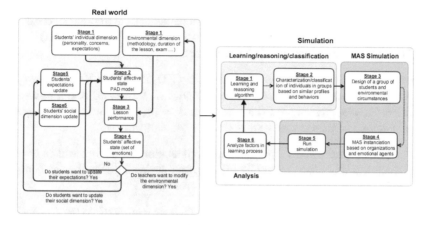

Fig. 1 Workflow process in the real-world and simulation.

in the environment modify students' affective state and expectations or the social dimension during the period of time of a lesson. The information collected in each lesson of several subjects during a course is stored and used as input for the *simulation*. This process for gathering information is repeated in each lesson[1].

Simulation starts by processing all the information collected from the real world. Using learning and reasoning techniques (i.e., machine-learning and case-based reasoning) individuals from the real learning environment are classified into groups of students with similar profiles and behaviors. Once a set of students categories has been extracted and the teacher has decided the environmental circumstances that s/he would like to test (i.e., type of methodology, tasks, group activity, etc.), the Multi-Agent System (MAS) is instantiated and the simulation starts. The MAS is based on the proposed student's affective model described in Section 3. After the simulation, results from the simulation are analyzed and used as input for the learning and reasoning process. In the meanwhile, new data obtained from students' experience during lessons in the real learning environment can be also considered as input for the learning and reasoning process that will provide feedback to the simulation.

4.1 Multi-Agent Architecture for Educational Environments

The main component of the simulation part is the Multi-Agent System (MAS). The proposed MAS is based on THOMAS [18] and the student's affective model described in Section 3. The proposed MAS is made up of a set of virtual organizations (VO) (i.e, group of students that are enrolled in a subject) and emotional autonomous agents (i.e., students) that are, at least, in one VO (i.e, they are at least enrolled in one subject).

---

[1] Note that, once the data is filled the first time, the students or the teacher do not necessarily require to repeat all the steps at each lesson

These agents interact and are influenced by other agents, environmental conditions, their individual characteristics, and the events that occur in the VO (subject).

Formally, the proposed MAS can be defined as follows:

DEFINITION 1 *(System). The system is a tuple (VO, A), where VO is a set of virtual organizations (i.e., subjects) $VO = \{VO_1, \ldots, VO_n\}$ and A is a finite set of emotional autonomous agents (i.e., students) $A = a_1, \ldots, a_n$, where each agent a_j should belong at least to one $VO_i \in VO$.*

A virtual organization in the proposed computational model considers information about learning goals, the profiles that appear during a lesson (i.e., students and teacher), and the actions associated to these profiles.

DEFINITION 2 *(Virtual Organization). The VO_i is defined as $(\mathcal{G}_i, SD_i, FD_i)$ where:*

- $\mathcal{G}_i = \{g_1, \ldots g_n\}$ *is the set of goals associated to the virtual organization. In the educational context, the goals of an organization correspond to improving learning factors such as satisfaction, percentage of comprehension, knowledge speed acquisition, and motivation.*
- $SD_i = \{E_i, OR_i, Relations\}$ *is the structural dimension that defines roles and relations among entities. E_i refers to the set of entities (virtual organizations, agents, or both) that are inside the organization. For instance, students that are enrolled in a subject (i.e., agents that belong to a VO_i) or a seminar that is inside a subject (i.e., VO_i inside a VO_k). $OR_i \in R$ refers to the roles that can be played inside the organization (i.e., teacher, student, group leader, etc.). Relations defines the relationship among roles inside the organization. The organizational topology in this context is hierarchical where the teacher is in the first level of the hierarchy and students in the second level. More roles can be defined and other structures considered.*
- FD_i *is the functional dimension. It describes the set of actions that agents can do. For instance, in the context of a lesson, the actions that a teacher may perform are: starting a new lesson, proposing activities, or establish a methodology among others. Actions related to students can be: assist to a lesson, participate, or take an exam among others.*

Considering the student's affective model described in section 3, we formally define the Emotional Agent.

DEFINITION 3 *(Emotional Agent). The emotional agent a_j represents a student and is characterized by a tuple (I_j, En_j, S_j, Aff_j) where:*

- $I_j = \{C_j, P_j, Ex_j, r_j\}$ *represents the individual dimension that consists of: C_j represents the concerns of the agent (i.e., get good marks, learning, pass all the exams, enjoy), P_j represents the personality of the agent (i.e., Big Five personality), Ex_j represents the expectations of the agent (i.e., finish the tasks, understand the new concepts of a lesson, end the lesson without doubts), r_j the role that the agent plays (i.e., teacher, student, spokesperson, etc.),*

- En_j represents the agent's knowledge about the environment where the agent is located (i.e., class duration, timetable, subject, breaks, learning methodology, lesson's number, tasks, exams' results, etc.),
- S_j represents the social dimension (i.e., classroom climate, student-peers relation, teacher-student relation),
- Aff_j represents the affective state in a temporal point as an aggregation of emotions related to the learning process.

DEFINITION 4 (Role) A role $r_\ell \in OR_i$ is defined by the tuple $(\phi_\ell, Act_{r_\ell})$, where:

- ϕ_ℓ is the role's name. In educational context the roles would be teacher and student.
- Act_{r_ℓ} is the set of actions associated to the role. Each action is defined by the tuple $(In_\ell, O_\ell, P_\ell, Eff_\ell)$ (i.e., Inputs, Outputs, Preconditions, and Effects). For instance, in an educational context, the actions associated to the role teacher could be: starting a new lesson, proposing activities, or establish a methodology.

The proposed computational model allow us to define a MAS where emotional agents (i.e., students) can interact autonomously in the context of a virtual organization (i.e., an specific subject). Each agent that plays the role *student* has a set of beliefs about itself, the environment, and the relationship with others. Agents have also a set of possible actions (i.e., participate, interact with peers, submit tasks, do an exam, etc.). These actions may imply a change in the beliefs of other agents with whom they interact. Moreover, in each virtual organization, there is an agent that plays the role *teacher*. This role has associated a set of actions that can update information about the environment (i.e., the lesson) such as create an individual activity, update methodology, or evaluate. When there is a change in the environment, agents are informed and they, based on their circumstances (i.e., their beliefs about their individual, environmental and social dimensions) can update their affective state accordingly, and, therefore update degree of achievement of the learning goals of the virtual organization. All these features of the computational approach facilitate the simulation of virtual learning environments where is possible to estimate the impact of schooling strategies over the learning process with the minimum intervention of students.

5 Conclusions

Emotions play an important role in learning environments. In this paper, we propose a model that considers a set of factors from a temporal multi-level perspective and events that influence on student's emotions. This temporal multi-level perspective considers three dimensions that include static and dynamic features: individual (i.e., concerns, personality, and expectations), environmental (i.e., features related to lesson's planning and organization), and social (i.e., classroom-climate, student-peers relation, teacher-student relation). This set of features and events that can occur during a lesson, influences on student's emotions, and therefore, in the expected learning factors such as satisfaction, knowledge acquisition, motivation, or comprehension. Based on this model for analyzing the factors that influence student's affective state, we propose a MAS architecture based on organizations that integrates emotional agents to incorporate the student's affective model and be able to simulate the effects of changes in the environment (i.e., the lesson) over the affective-state of students. We also present the workflow for the inclusion of information collected about these effects over the student's affective state in real and simulated environments. As future work, we plan to validate our proposal by applying it in a real educational environment.

Acknowledgements This work is supported by the following national projects TIN2015-65515-C4-1-R, TIN2014-55206-R and PROMETEOII/2013/019, and the European project H2020-ICT-2015-688095.

References

1. Pekrun, R., Frenzel, A.C., Götz, T., Perry, R.P.: The control-value theory of achievement emotions: an integrative approach to emotions in education. In: Emotion in education, pp. 13–36. Academic Press (2007)
2. Mehrabian, A.: Pleasure-arousal-dominance: A general framework for describing and measuring individual differences in Temperament. Current Psychology **14**(4), 261–292 (1996)
3. Ryckman, R.M.: Theories of Personality. PSY 235 Theories of Personality Series. Thomson/Wadsworth (2007)
4. Lazarus, R.: Emotion and Adaptation. Oxford University Press (1994)
5. Ortony, A., Clore, G.L., Collins, A.: The Cognitive Structure of Emotions. Cambridge University Press, July 1988
6. Scherer, K.R.: Appraisal considered as a process of multilevel sequential checking. Appraisal Processes in Emotion: Theory, Methods, Research **92**, 120 (2001)
7. Chalfoun, P., Chaffar, S., Frasson, C.: Predicting the emotional reaction of the learner with a machine learning technique. In: ITS 2006 (2006)
8. Chaffar, S., Cepeda, G., Frasson, C.: Predicting the Learner's Emotional Reaction towards the Tutor's Intervention. In: ICALT, pp. 639–641 (2007)
9. Fatahi, S., Kazemifard, M., Ghasem-Aghaee, N.: Design and implementation of an e-learning model by considering learner's personality and emotions. In: Advances in Electrical Engineering and Computational Science, vol. 39, pp. 423–434 (2009)
10. Ibrahimoglu, N., Unaldi, I., Samancioglu, M., Baglibel, M.: The relationship between personality traits and learning styles: A cluster analysis. Asian Journal of Management Sciences and Education **2**(3) (2013)
11. Costa, P.T., McCrae, R.R.: Four ways five factors are basic. Personality and Individual Differences **13**(6), 653–665 (1992)
12. Sander, P., Stevenson, K., King, M., Coates, D.: University students' expectations of teaching. Studies in Higher Education **25**(3), 309–323 (2000)
13. Voss, R., Gruber, T., Szmigin, I.: Service quality in higher education: The role of student expectations. Journal of Business Research **60**(9), 949–959 (2007)
14. Schunk, D., Pintrich, P., Meece, J.: Motivation in Education: Theory, Research, and Applications. Pearson/Merrill Prentice Hall (2008)
15. Graham, S.: A review of attribution theory in achievement contexts. Educational Psychology Review **3**(1), 5–39 (1991)
16. Wang, M.C., Haertel, G.D., Walberg, H.J.: What Helps Students Learn? Spotlight on Student Success (1997)
17. Meyer, D.K., Turner, J.C.: Re-conceptualizing emotion and motivation to learn in classroom contexts. Educational Psychology Review **18**(4), 377–390 (2006)
18. Del Val, E., Criado, N., Carrascosa, C., Julian, V., Rebollo, M., Argente, E., Botti, V.: THOMAS: a service-oriented framework for virtual organizations. In: 9th AAMAS, pp. 1631–1632 (2010)

A General Framework for Testing Different Student Team Formation Strategies

Juan M. Alberola, Elena del Val, Victor Sanchez-Anguix and Vicente Julián

Abstract One of the most important problems faced by teachers is grouping students into proper teams. The task is complex, as many technical and interpersonal factors could affect team dynamics, with no clear indication of which factors may be more relevant. Not only the problem is conceptually complex, but its computational complexity is also exponential, which precludes teachers from optimally applying strategies by hand. The tool presented in this paper aims to cover both gaps: first, it provides a range of grouping strategies for testing, and second, it provides artificial intelligence mechanisms that in practice tone down the computational cost of the problem.

Keywords Team formation · Teamwork · Education · Belbin · Myers-briggs

1 Introduction

A recent graduate has just been hired by an e-commerce company after finishing her degree in Computer Science. In the first day in the job, she is introduced to her team, which is composed by a variety of profiles. Let us be honest, it did not matter if the student had finished a degree in Computer Science, Statistics, Literature, or Ecology. It also did not matter whether she was going to be part of an e-commerce company, a manufacturing company, a bank, or a retail shop. The scenario will be in all cases similar, as the problems faced by these organizations will be so complex that they will require the action of teams [13, 15].

J.M. Alberola(✉) · E. del Val · V. Julián
Universitat Politècnica de València, Valencia, Spain
e-mail: {jalberola,edelval}@dsic.upv.es

V. Sanchez-Anguix
Department of Computing, Coventry University, Gulson Road, Coventry CV1 2JH, UK
e-mail: ac0872@coventry.ac.uk

© Springer International Publishing Switzerland 2016   23
M. Caporuscio et al. (eds.), *mis4TEL*,
Advances in Intelligent Systems and Computing 478,
DOI: 10.1007/978-3-319-40165-2_3

Despite the ability for teams to tackle complex problems, it has also been documented that teamwork is by no means a trivial soft skill and it requires to be polished. If team dynamics are not adequate, the results can be totally opposite to those intended when the team was formed [3, 6]. Given these circumstances, teamwork has been introduced in Higher Education as one of the core general competences in the course programmes [4, 14]. Nevertheless, a very unsatisfactory team experience may preclude team members from adequately polishing their teamwork skills, and it may overly emphasize the negative experiences in future scenarios. Therefore, for students to correctly sharpen their team skills, it is necessary to create an adequate environment in educational settings.

The aforementioned problem has no straightforward solution. First of all, there are many criteria that could affect team dynamics inside a team: technical skills, soft skills, knowledge distribution, personality, friendship, common interests, goals' compatibility, languages, schedule's compatibility, and so forth. As the reader can imagine, the range of possible grouping strategies is almost infinite and there is not a clear indication of which strategies may outperform others in terms of facilitating teamwork. Secondly, even if the best grouping criteria was clear, the problem of optimally allocating students into groups is exponential with the number of students, which makes the process of obtaining the fitness of each possible combination, difficult to be calculated by hand. Therefore, computational tools may be necessary to provide support to lecturers in that task.

The tool that we present in this paper aims to cover such gap. First of all, lecturers can employ multiple defined grouping strategies to test with their own students. Second, the tool applies artificial intelligence optimization methods to provide an optimal solution for the grouping problem, which tones down the exponential cost in practice. The tool is an evolution of our previous work [1, 16], where we offered a grouping method for students based on Belbin's role taxonomy. The remainder of this paper is organized as follows. First, Section 2 describes the proposal in detail. Section 3 shows how the proposed tool has been applied to test several team formation strategies in a real educational environment. Finally, Section 4 presents some concluding remarks and future work.

2 Framework Design

In this section, we describe in detail the team formation tool. The tool's core components can be grouped into three different functionalities: information gathering, team formation, and feedback and analysis.

2.1 Information Gathering

All team formation activities to be supported by the tool always start with some information gathering. To begin with, all the users (lecturer and students) must be registered and logged into the system. The first time that a specific team activity is created, the teacher needs to specify some required information about the activity (e.g. description, start date, end date, etc.).

From the students perspective, some extra information may be required depending on the team formation strategy. Thus, some initial questionnaires may be prepared to be filled by the students. The current version of the framework provides the Belbin Self-Perception Inventory for those strategies based on Belbin's model [7] and the Myers-Briggs Type Indicator (MBTI) personality test for the strategy based on the MBTI model [8]. The former is composed by several questions related to the own perception when working with others: the contribution, conflict resolution, task satisfaction, etc. The later is composed by questions related to everyday situations in which the student has to select between two options. These approaches have been widely applied in educational environments, showing the benefits of the results [2, 10, 11, 17, 18, 19].

Apart from these well-known questionnaires, some information about the students can also be introduced such as their academic performance (e.g. marks), their main motivations (to obtain high performance, to learn, etc.) or their relationship with other students and with the teacher. The tool may also use information from past team activities registered in the systems.

2.2 Team Formation

The information gathered in the previous component, and any other information already registered in the system is then used for creating teams in the recently created activity. For that, the lecturer should select a team formation strategy and define general parameters such as the maximum and minimum size for the teams, or any incompatibility among students (e.g., personal conflicts).

Currently, the tool supports a variety of team formation strategies: random teams, a strategy based on the students' performance, a strategy based on the students' concerns, two strategies based on Belbin's role taxonomy (by using self-perception and peers' feedback), and one strategy based on the MBTI model.

The problem of grouping students into optimal teams is equivalent to a coalition structure generation problem. The problem consists of partitioning elements from a global set into exhaustive and disjoint groups, so that the global benefits of the system are optimized. In our problem, the elements of the set are the students participating in the team activity. Let $A = \{a_i, ..., a_n\}$ be a set of students and let $T_j \in A$ be a subset of A called *team*, the expected value of a team T_j is given by a characteristic function $v(T_j) : 2^A \rightarrow \mathbb{R}$ that assigns a real-valued payoff to each team T_j. This expected value depends on the specific team formation strategy that is selected.

A *team structure* $S = \{T_1, T_2, ..., T_k\}$ is a partition of teams such that $\forall i$, $j(i \neq j), T_j \cap T_i = \emptyset, \bigcup_{\forall T_j \in S} T_j = A$. The expected value of a team structure is denoted by $v(S)$, which is an evaluation function for the team structure. In this work, we consider that the quality of each team is independent of other teams. Therefore, we can calculate the value of the team structure as $v(S) = \sum_{T_j \in S} v(T_j)$.

The goal of the team formation problem is to determine an optimal team structure for the classroom $\underset{S \in 2^A}{argmax} \ v(S)$. Taking into account the specific equation, which is associated to each team formation strategy, to calculate the expected value of a given team, our team formation problem requires to solve the following expression:

$$\underset{S \in 2^A}{argmax} \sum_{T_j \in S} v(T_j) \tag{1}$$

It turns out that partitioning a set of students into disjoint teams while optimizing a social welfare function corresponds to the formalization of coalition structure generation problems. In order to solve this problem, we formally define the coalition structure generation problem as a linear programming problem [9] and solve it with the commercial software *ILOG CPLEX 12.5*[1]. Next, we define in general terms each of the grouping strategies.

Random Strategy

The random team formation strategy is mainly provided for those scenarios in which no information from students is available (cold start). Therefore, the algorithm associated to this strategy provides randomly generated teams, which may or may not be well-balanced.

Strategy Based on Student Concerns

This team formation strategy is focused on grouping students according to similar learning concerns. Concerns represent the individual achievement goals of the student in the long term. In the learning context, we focus on what students hope to accomplish during the course (i.e., learn as much as they can about a subject, personal satisfaction, doing the minimum effort to pass the course, or doing well on exams to obtain a high mark). These concerns can influence on how students approach, experience, and perform in classes [5].

---

[1] http://www.ibm.com/software/commerce/optimization/cplex-optimizer/

Strategy Based on Academic Performance

This team formation strategy is purely based on the academic performance obtained by the students. In this approach, students are grouped according to their marks. The current implementation obtains teams whose students have the most similar academic marks (i.e. the standard deviation of their marks is minimized).

Strategy Based on Belbin Self-Perception

This team formation strategy is based on one of the most important theories regarding successful team dynamics, which is Belbin's role taxonomy [7]. Belbin identifies eight heterogeneous behavioral patterns (or roles) that are present in many successful teams in the industry: *plant (PL), resource investigator (RI), coordinator (CO), shaper (SH), monitor evaluator (ME), team worker (TW), implementer (IM), and completer finisher (CF)*. These roles should be played by the different team members in order to facilitate successful teamwork. This model has been widely applied to the classroom environment, where it has shown a variety of benefits [11, 17, 19]. In this approach, each student has one or several roles with a significant score according to the Belbin Self-Perception Inventory. Thus, the expected value of a team is calculated according to the role diversity within the team.

Strategy Based on Belbin and Peers' Feedback

The previous strategy may have an important shortcoming. It is based on one's self-perception, which many times may differ from the behaviour that we actually display in a team. A more educated guess on one's behavior may be to include the feedback of those that have worked with the individual. This strategy relies on that assumption.

   After each activity, each student evaluates his/her peers by stating the most predominant role of each of his/her teammates. Then, the strategy calculates an estimation of the expected value of the team given the history of evaluations received for each student. After each team activity, the evaluation history grows and, therefore, a more appropriate estimation of students' predominant roles should be present. In order to tackle uncertainty in information provided by students, the strategy employs Bayesian learning, and a probability distribution over roles is calculated for each student. A more detailed description of this strategy can be found in [1].

Strategy Based on MBTI

The MBTI [8] model focuses on measuring the world's perception and the decision-making process of an individual, from a psychological point of view. The questionnaire defines four different dichotomies of human preferences, each divided in two opposite dimensions: Extroversion (E) -Introversion (I); Sensing (S) - Intuition (N);

Thinking (T) - Feeling (F); and Judging (J) - Perceiving (P). This indicator has also been widely used for team formation in educational environment [2, 10, 18]. The basis for team formation in this approach is focused on balancing the personality types. In the current implementation, we calculate the team diversity by using the formula proposed by Pieterse et al. [12], which measures the degree of diversity for each preference dimension within a team. The algorithm associated to this team formation strategy obtains teams in which the number of students with preference dimensions are balanced in as many as the four dichotomies as possible.

2.3   Feedback and Analysis

After the completion of an activity, the framework provides some tools to provide feedback from the participants, and to analyze the results.

On the one hand, some questionnaires are prepared to evaluate the teamwork performance in terms of several aspects such as the individual's satisfaction with the team, the individual's satisfaction with each partner, the individual's satisfaction with the work, team dynamics, and the lecturer's opinion.

This information is processed and delivered to the teacher in order to analyze team performance, and to compare the activity results with previous activities with different/similar team formation strategies.

3   Case Study

We applied this framework in a real education environment, in order to test different team formation strategies. We carried out this in a first semester (2015/2016) module pertaining to the *Tourism Degree Program of* at *Universitat Politècnica de València*. Specifically, following we show the results of the Belbin self perception strategy, and using the MBTI strategy. It should be highlighted that these evaluations were solely carried out for testing purposes and check the behavior of students in the classroom. More specifically, we were interested in studying whether or not well-balanced teams could be formed in terms of the different heuristics. In no case this evaluation is aimed to compare the performance of the different strategies. A total of 46 students participated in the test. We employed two student cohorts: Group A with 30 students, and Group B with 16 students. According to this, we created teams of 5 members in Group A and teams of 4 members in Group B. At the beginning of the course, students filled the Belbin Self-Perception Inventory and the Myers-Briggs Type Indicator.

We conducted a first test by forming teams according to the Belbin self perception strategy. The results of the Belbin Self-Perception Inventory provide for each student a score indicating the predominance of each role. The performance of a team is measured on the premise that each role should have at least one significant score

Table 1 Number of students with significant scores in the different Belbin roles

	CW	CH	SH	PL	RI	ME	TW	CF
Significant scores	2	11	15	16	4	5	10	21

Table 2 Students distributed according to their preference personality dimension

	E	I	N	S	F	T	P	J
Group A	19	4	15	11	14	9	4	20
Group B	10	2	10	5	11	4	6	7
Total	29	6	25	16	25	13	10	27

among team members' scores. Thus, the highest expected value for a team is the maximum number of roles that can be played by the whole team, i.e. $v(S) = 8$. The role distribution is shown in Table 1. It can be observed, that there were roles such as CW that were played by only a few students, while other roles such as CF were played by a larger number of students. Despite this, the average expected value of the teams formed by the framework were 6.17 of a maximum of 8 in Group A and 5 of a maximum of 8 in Group B. Therefore, in both cases it was possible to form heterogeneous teams that maximize the team formation strategy.

The second test focused on forming teams according to the MBTI strategy. The results of the Myers-Briggs Type Indicator associate a dimension preference for each student. Table 2 shows the number of students distributed according to their preference for the eight personality dimensions defined by the MBTI model. We must note that those students who had not shown a clear preference in some dichotomy, were counted as having preference in both dimensions of this dichotomy. According to this strategy, the highest expected value for a team is the number of personality dimensions that are defined, i.e. $v(S) = 8$. Similar to the previous test, the distribution of personalities among the students was not balanced. As an example, only 6 students had preference in the I dimension while 29 students had preference in the E dimension. Despite this, the average expected value of the teams was very good compared to the optimal, being 5.65 of a maximum of 8 in Group A and in and 6 of a maximum of 8 in Group B. Again, in both cases it was possible to form heterogeneous teams that maximize the team formation strategy.

This evaluation allowed us to test the applicability of these strategies in a real class environment. It was important to assess whether balanced teams would come from a real student population, and the results support this. We are currently running an additional test with these strategies that will allow us to compare both in terms of team performance.

4 Conclusions

In this article, we have presented a framework for team formation in educational environments. The framework provides different functionalities to lecturers and students that support the process of team formation. Specially, the framework includes different strategies for team formation, allowing the lecturer to select the most suitable approach depending on the group of students. On top of that, the tool also allows lecturers to compare the results of different team formation strategies after the completion of team activities. We carried out a case study in the classroom to assess the applicability of some of these strategies in a real scenario. The results show that despite the fact that the distribution of roles/personalities is not homogeneous, resulting teams are well-balanced in terms of the maximum value achievable by the heuristics. The initial perception of some students is not very optimistic since some of them would prefer to work with their friends rather than others. However, as they start working inside the teamwork, their opinions tend to change. As for future lines of research, we plan to compare the performance of teams formed by different strategies, in terms of the activities that they carry out.

Acknowledgements This work is supported by the following national projects TIN2015-65515-C4-1-R, TIN2014-55206-R and PROMETEOII/2013/019, and the european project H2020-ICT-2015-688095.

References

1. Alberola, J.M., del Val, E., Sanchez-Anguix, V., Julian, V.: Simulating a collective intelligence approach to student team formation. In: Hybrid Artificial Intelligent Systems, pp. 161–170. Springer (2013)
2. Amato, C.H., Amato, L.H.: Enhancing student team effectiveness: Application of myers-briggs personality assessment in business courses. Journal of Marketing Education **27**(1), 41–51 (2005)
3. Behfar, K., Friedman, R., Brett, J.M.: The team negotiation challenge: Defining and managing the internal challenges of negotiating teams. In: IACM 21st Annual Conference Paper (2008)
4. Dunne, E., Rawlins, M.: Bridging the gap between industry and higher education: Training academics to promote student teamwork. Innovations in Education and Teaching International **37**(4), 361–371 (2000)
5. Harackiewicz, J.M., Barron, K.E., Tauer, J.M., Carter, S.M., Elliot, A.J.: Short-term and long-term consequences of achievement goals: Predicting interest and performance over time. Journal of Educational Psychology **92**(2), 316 (2000)
6. Kerr, N.L., Tindale, R.S.: Group performance and decision making. Annu. Rev. Psychol. **55**, 623–655 (2004)
7. Meredith Belbin, R.: Management teams: Why they succeed or fail. Human Resource Management International Digest **19**(3) (2011)
8. Myers, I.B.: The myers-briggs type indicator: Manual (1962)
9. Ohta, N., Conitzer, V., Ichimura, R., Sakurai, Y., Iwasaki, A., Yokoo, M.: Coalition structure generation utilizing compact characteristic function representations. In: Principles and Practice of Constraint Programming - CP 2009, vol. 5732, pp. 623–638 (2009)
10. Omar, M., Syed-Abdullah, S.-L., Hussin, N.M.: Analyzing personality types to predict team performance. In: Proceedings of the 2010 International Conference on Science and Social Research (CSSR), pp. 624–628 (2010)

11. Ounnas, A., Davis, H.C., Millard, D.E.: A framework for semantic group formation in education. Journal of Educational Technology & Society **12**(4), 43–55 (2009)
12. Pieterse, V., Kourie, D.G., Sonnekus, I.P.: Software engineering team diversity and performance. In: Proc. of SAICSIT 2006, pp. 180–186. South African Institute for Computer Scientists and Information Technologists (2006)
13. Ratcheva, V.: Integrating diverse knowledge through boundary spanning processes-the case of multidisciplinary project teams. International Journal of Project Management **27**(3), 206–215 (2009)
14. Sancho-Thomas, P., Fuentes-Fernández, R., Fernández-Manjón, B.: Learning teamwork skills in university programming courses. Computers & Education **53**(2), 517–531 (2009)
15. Tarricone, P., Luca, J.: Employees, teamwork and social interdependence-a formula for successful business? Team Performance Management: An International Journal **8**(3/4), 54–59 (2002)
16. Val, E.D., Alberola, J.M., Sánchez-Anguix, V., Palomares, A., Teruel, M.D.: A team formation tool for educational environments. In: Trends in Prac. Apps. of Heterogeneous Multi-Agent Systems, vol. 293, pp. 173–181 (2014)
17. van de Water, H., Ahaus, K., Rozier, R.: Team roles, team balance and performance. Journal of Management Development **27**(5), 499–512 (2008)
18. Varvel, T., Adams, S.G., Pridie, S.J., Ruiz, B.C.: Ulloa. Team effectiveness and individual myers-briggs personality dimensions. Journal of Management in Engineering **20**(4), 141–146 (2004)
19. Yannibelli, V., Amandi, A.: A deterministic crowding evolutionary algorithm to form learning teams in a collaborative learning context. Expert Systems with Applications **39**(10), 8584–8592 (2012)

Monitoring Level Attention Approach in Learning Activities

Dalila Durães, Amparo Jiménez, Javier Bajo and Paulo Novais

Abstract In this article we focus on a new field of application of ICT techniques and technologies in learning activities. With these activities with computer platforms, attention allows us to break down the problem of understanding a speculative scenario into a series of computationally less demanding and localized lack of attention. The system considers the students' attention level while performing a task in learning activities. The goal is to propose an archi-tecture that measures the level of attentiveness in real scenario, and detect patterns of behavior in different attention levels among different students. Measurements of attention level are obtained by a proposed model, and user for training a decision support system that in a real scenario makes recommendations for the teachers so as to prevent undesirable behavior.

Keywords Learning activities · Attention level · Performance

1 Introduction

Teaching should be solidly grounded to the absolute understanding of how the process of learning occurs, so that instructional strategies could be efficient and lead to persistent knowledge. When students use technologies in learning activities

D. Durães(✉) · J. Bajo
Department of Artificial Intelligence, Technical University of Madrid, Madrid, Spain
e-mail: d.alves@alumnos.upm.es, jbajo@fi.upm.es

A. Jiménez
Faculty of Computer Science, Pontifical University of Salamanca, Salamanca, Spain
e-mail: ajimenezvi@upsa.es

P. Novais
Algoritmi Center, Minho University, Braga, Portugal
e-mail: pjon@di.uminho.pt

© Springer International Publishing Switzerland 2016
M. Caporuscio et al. (eds.), *mis4TEL*,
Advances in Intelligent Systems and Computing 478,
DOI: 10.1007/978-3-319-40165-2_4

distractions might be occurs with other applications and the acquisition of knowledge can't occur. It's crucial to improve the learning process and to mitigate problems that might occur in an environment with learning technologies. Learning theories provide insight into the very complex processes and factors that influence learning and provide precious information to be used to design instruction that will produce prime results. Besides, each student has its own particular way of assimilating knowledge, that is, his learning style. Learning styles specify a student's own way of learning. Someone that has a specific learning style can have difficulties when submitted to another learning style [1]. When the given instruction style matches the student's learning style, the process is maximized which guarantees that the student learns more and more easily.

Technologies that enhance learning environments are ideal for generating learning style-based instructional material in large classes, as they don't have the same limitations as human instructors due to the lack of resources and time to focus on individual students. With this recommendation the teacher can improve some strategies that may increase the level of attention and engagement of the students and they might improve learning.

In this article we focus on a new field of application of ICT techniques and technologies in learning activities. The goal is to propose an architecture aimed at capturing and measuring the level of students' attentiveness in real scenarios and dynamically provide recommendations to the teacher in order to improve the better learning styles for each student.

2   Theoretical Foundations

The concept of attention is commonly used either to describe the active selection of information from the environment or the processing of information from internal sources [2]. It also can be defined as filtering input space to more important spaces in processing. Attention means focusing on thought clearly, among one of several subjects or objects that may capture mind simultaneously. Attention implies the concentration of mental powers upon an object by close or careful observing or listening, which is the ability or power to concentrate mentally. Attention means to cut things to deal effectively to other things. The level of the learner's attention affects learning results. The lack of attention can define the success of a student. In learning activities, attention is also very important to perform these tasks in an efficient and adequate way. In learning activities with computer platforms, computational attention allows us to break down the problem of understanding a speculative scenario into a series of computationally less demanding with visual, audio, and linguistic approach [3].

Being a cognitive process, attention is strongly connected with learning [4]. When it comes to acquiring new knowledge, attention can be considered one the most important mechanisms [5]. The degree of the learner's attention affects learning results. The lack of attention can define the success of a student and in learning activities, attention is very important in order to perform these tasks in an efficient and adequate way.

Generally, there are some factors that influence attention level: stress, mental fatigue, and anxiety.

2.1 Stress

When students are subjected to increasing periods of work with a progressive focus on autonomy and continuous assessment, the workload is perceived as stressful and usually leads to emotional disorders, which affects attention and concentration [6]. However, in small periods of time, stress tends to behave in a more efficient way, decreasing the number of unnecessary actions as students are more focused on their tasks [4].

Human stress is a state of tension that is created when a person responds to demands and pressures [7]. However students react in different ways, whereas a situation might be stressful for a student and relaxing for another. Students aren't affected exactly the same way or suffer from the same degree of stress. Although sooner or later in life one goes through a stressful situation [8]. When the students are forced to a higher number of tasks and assessment, they have to set priorities and the level of fear increases, because they don't want to fail. Consequently, the level of pressure increases causing stress.

2.2 Mental Fatigue

Usually, the term mental fatigue is a cognitive ability that is decreased and used to describe a sequence of manifestations like lack of concentration, loss of attention, and slower reaction in response time. When students are working for an extended period of time, they often end up feeling the effect of inattention, reflected in impaired task performance and reduced engagement to continue working [9, 10]. In addition, one student that feels lower performance also has a harder time concentrating, getting easily distracted [11, 12], an indication that mental fatigue can have effects on selective attention.

Mental Fatigue can occur at any time during the day. Depending on its duration and intensity, mental fatigue can make the carrying out of daily tasks increasingly hard or even impossible [6]. Learning is one of the functions that become impaired when under fatigue. The importance of addressing this issue when students are using learning activities is very important for a teacher. Teachers need to be sensible to the state of mind of their students, impairing their ability to adapt both the contents and the teaching strategy accordingly [13].

2.3 Anxiety

Anxiety is an aversive emotional and motivational state occurring in threatening circumstances. Generally, anxiety has an adverse effect on attention because it causes inattention [14]. When, in a small period of time, it leads to compensatory strategies like enhanced effort. Anxiety can't change the level of attention [15].

If we consider efficiency the relation between success and the resources spent on a task, anxiety is meant to have a negative influence in the field of cognition and attention through its cognitive interference by preempting the processing and temporary storage capacity of working memory.

3   A Dynamic Approach to Monitor Attention

When students are affected by positive or negative states, they produce different kinds of thinking and this might hold important implications on the educational and training perspective. This means that students who are caught in affective

Table 1 Data acquisition features.

Symbol	Feature	Description
		Mouse Events
mv	Mouse Velocity	The distance travelled by the mouse (in pixels) over the time (in milliseconds).
ma	Mouse Aceleration	The velocity of the mouse (in pixels/milliseconds) over the time (in milliseconds).
cd	Click Duration	the timespan between MOUSE_UP events, whenever this timespan is inferior to 200 milliseconds.
tbc	Time Between Clicks	the timespan between two consecutive MOUSE_UP and MOUSE_DOWN events, i.e., how long did it took the individual to perform another click.
dbc	Distance Between Click	represents the total distance travelled by the mouse between two consecutive clicks, i.e., between each two consecutive MOUSE_UP and MOUSE_DOWN events.
ddc	Duration Distance Clicks	the time between consecutive MOUSE_UP and MOUSE_DOWN events.
edbc	Excess Distance Between Clicks	represents the excess total distance travelled by the mouse between two consecutive clicks, i.e., between each two consecutive MOUSE_UP and MOUSE_DOWN events.
aedbc	Absolute Excess Distance Between Click	this feature measures the average distance of the excess total distance travelled by the mouse between two consecutive clicks, i.e., between each two consecutive MOUSE_UP and MOUSE_DOWN events.
asdbc	Absolute Sum Distance Between Clicks	this feature measures the average sum of distance that the mouse travelled between each two consecutive MOUSE_UP and MOUSE_DOWN events.
dplbc	Distance Point to Line Between Clicks	this feature will compute the distance between two consecutive MOUSE_UP and MOUSE_DOWN events.
adpbc	Absolute Distance Point Between Clicks	this feature will compute the average distance between two consecutive MOUSE_UP and MOUSE_DOWN events.
		Keyboard Events
kdt	Key Down Time	the timespan between two consecutive KEY_DOWN and KEY_UP events.
tbk	Time Between Keys	the timespan between two consecutive KEY_UP and KEY_DOWN events
kdtv	Key Down Time Velocity	The times that two consecutive keys are press

states such as anger or depression do not process and engage information efficiently. When that occurs it would be important to be able to notify the teacher, so he can be able to dynamically modify the teaching style according students' feedback signals which include cognitive, emotional and motivational aspects.

The first stage of the proposed system is data collection, which was designed and carried out using a logger application developed in previous work [5]. The data collected by the logger application characterizing the students' interaction patterns is aggregated in a server to which the logger application connects after the student logs in. This application runs in the background, which makes the data acquisition process, a completely transparent one from the point of view of the student.

To monitor students' attention level, a log tool was developed, logging some features regarding student-computer interaction through particular operating system events read from the use of the computer's mouse and keyboard. Table 1 summarizes these events.

It possible to collected data that describes the interaction with both the mouse and the keyboard [13]. Previous work on this data collection tool and analysis can be found in [6] where a deeper analysis about this process is explained in detail.

4   Proposed Monitoring Architecture

From the features presented in the previous section, we can conclude that it is possible to obtain a measure of the students' attention level. Once information about the individual's attention exists in these terms, it is possible to start monitoring attentiveness in real-time and without the need for any explicit or conscious interaction. This makes this approach especially suited to be used in learning activities in which students use computers, as it requires no change in their working routines. This is the main advantage of this work, especially when compared to more traditional approaches that still rely on questionnaires (with issues concerning wording or question construction), special hardware (that has additional costs and is frequently intrusive) or the availability of human experts.

Figure 1 depicts the process through which the system operates; it is possible to observe the different classifications of information in order to allow, in the end, the management of attention level.

4.1   Dynamic Student Monitoring Architecture

While the student conscientiously interacts with the system and takes his/her decisions and actions, a parallel and transparent process take place in which the Dynamic Student Monitoring Architecture uses this information. This module, upon converting the sensory information into useful data, allows for a contextualized analysis of the operational data of the students. This framework performs this

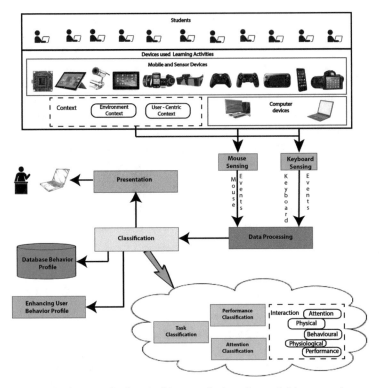

Fig. 1 Dynamic Student Monitoring Architecture for learning activities scenarios.

contextualized analysis. Then, the student's profile is updated with new data, and the teacher receives feedback from this module.

The system developed to acquire data from normal working compiles information from students' learning activities with mouse and keyboard which act as sensors. The proposed framework includes not only the complete acquisition and classification of the data, but also a presentation level that will support the human-based or autonomous decision-making mechanisms that are now being implemented. It is a layered architecture.

The Mouse and Keyboard Sensing layers are charged for capturing information describing the behavioral patterns of the students', and receiving data from events mouse and keyword students'. This layer encodes each event with the corresponding necessary information (e.g. timestamp, coordinates, type of click, key pressed). These data are further processed, stored and then used to calculate the values of the behavioral biometrics. Mouse movements can also help predict the state of mind of the user, as well as keyboard usage patterns.

The Data Processing layer is responsible to process the data received from the Data Acquisition layer in order to be evaluate those data according to the metrics presented. It's important that in this process some values should be filtered to eliminate possible negative effects on the analysis (e.g. a key pressed for more than a certain amount of time). The system receives this information in real-time and calculates, at regular intervals, an estimation of the general level of performance and attention of each student.

The Classification layer is where the indicators are interpreted for example: interpreting data from the attentiveness indicators and to build the meta data that will support decision-making. When the system has an enough large dataset that allows making classifications with precision, it will classify the inputs received into different attention levels in real-time. This layer has access to the current and historical state of the group from a global perspective, but can also refer to each student individually.

For that, this layer uses the machine learning mechanisms. After the classification, the Enhancing User Behavior Profile layer is responsible for providing access to the lower layer. The Database Behavior Profile is also a very important aspect to have control off. This possibility allows to analyses within longer time frames. This layer, whose function detect student's mood preserving those information (actual and past) in the mood database. This information will be used by another sub-module, the affective adaptive agent, to provide relevant information to the platform and to the mentioned personalization module.

Finally at the top, the Presentation layer includes the mechanisms to build intuitive and visual representations of the attentiveness states of the students', abstracting from the complexity of the data level where they are positioned. At this point, the system can start to be used by the people involved, especially the teacher who can better adapt and personalize his teaching strategies. The actual students' mood information are displayed in the Presentation layer, and can be used to personalize instruction according to the specific student, enabling Teacher to act differently with different students, and also to act differently to the same student, according to his/her past and present mood.

5   Conclusions and Future Work

Technology make possible the enhanced of learning/teaching processes, overcoming restrictions such as qualified instructor's availability, time restrictions, and individual monitoring for instance. A framework is proposed to address these issues, especially to monitoring students in learning activities. Narrowing the scope of the study, a model to detect attentiveness is proposed, through the use of a developed log tool. With this tool it is possible to detect those factors dynamically and non-intrusively, making it possible to foresee negative situations, and taking actions to mitigate them. The door is then open to intelligent platforms that allow to analyze students' profile, taking into account their individual characteristics, and to propose new strategies and actions, minimizing issues such as stress, anxiety, and new environments, which can influence students' results and are closely related to the occurrence of conflicts. Moreover its possible maximized performance and attentiveness since the teacher is informed to the behavioral of each student. Enlarge this study to the use of smartphones and tablets, taking advantage of their new features such as several incorporated sensors, and high resolution cameras, is the next step that possible will allow a wider characterization of the student, making it possible to enhance learning experience, though better recommendation and personalization.

Acknowledgements This work has been supported by COMPETE: POCI-01-0145-FEDER-007043 and FCT – Fundação para a Ciência e Tecnologia within the Project Scope: UID/CEC/00319/2013.

References

1. Smith, L.H., Renzulli, J.S.: Learning style preferences: A practical approach for classroom teachers. Theory into Practice **23**(1), 44–50 (1984). doi:10.1080/00405848409543088
2. James, W.: The principles of Psychology – Part 1. Read Books, Ltd. pp. 220–270 (2013)
3. Yu-Fei M., Hong J.: Contrast-based image attention analysis by using fuzzy growing. In: ACM Multimedia, pp. 374–381 (2003)
4. Rodrigues, M., Gonçalves, S., Carneiro, D., Novais P., Fdez-Riverola, F.: Keystrokes and clicks: measuring stress on E-learning students. In: Management Intelligent System, vol. 220, pp. 119–126 (2013)
5. Pimenta, A., Carneiro, D., Neves, J., Novais, P.: A Neural Network to Classify Fatigue from Human-Computer Interaction. Neurocomputing **172**, 413–426 (2015)
6. Carneiro, D., Novais, P., Pêgo, J.M., Sousa, N., Neves, J.: Using mouse dynamics to assess during online exams. In: Hybrid Artificial Intelligent Systems, vol. 9121, pp. 345–356 (2015)
7. Gardell, B.: Worker Participation and Autonomy: a Multilevel Approach to Democracy at the Workplace. International Journal of Health Services **4**, 527–558 (1982)
8. Rodrigues, M., Riverola, F., Novais, P.: Na approach to Assess Stress in E-learning Students (2012)
9. Meijman, T.F.: Mental Fatigue and the Efficiency of Information Processing in Relation to Work Times. International Journal of Industrial Ergonomics **20**(1), 31–38 (1997)
10. Bartlett, F.C.: Ferrier Lecture: Fatigue Following Highly Skilled Work. Proceedings of the Royal Society of London. Series B-Biological Sciences **131**(863), 247–257 (1943)
11. Faber, L.G., Maurits, N.M., Lorist, M.M.: Menatl Fatigue Affects Visual Selective Attention. PloS One **7**(10), e48073 (2012)
12. Lorist, M.M., Kleim, M., Nieuwenhuis, S., Jong, R., Mulder, G., Meijman, T.F.: Mental Fatigue and Task Control: Planning and Preparation. Psychophysiology **37**(5), 614–625 (2000)
13. Pimenta, A., Gonçalves, S., Carneiro, D., Fde-Riverola, F., Novais, P.: Mental Workload Management as a Tool in e-learning Scenarios (2015)
14. Van der Linden, D., Frese, M., Meijman, T.F.: Mental Fatigue and the Control of Cognitive Processes: Effects on Perseveration and Planning. Acta Psychologica **113**(1), 45–65 (2003)
15. Eysenck, M.W., Derakshan, N., Santos, R., Calvo, M.G.: American Psychological Association **7**(2), 336–353 (2007)
16. Trigwell, K., Ellis, R., Han, F.: Relations between students' approaches to learning, experienced emotions and outcomes of learning. Studies in Higher Education **37**(7), 811–824 (2012). doi:10.1080/03075079.2010.549220

Guidelines for Participatory Design of Digital Games in Primary School

Gabriella Dodero and Alessandra Melonio

Abstract Game design with children with a participatory approach has been receiving an increasing attention in recent years. However, game design as a complex
design process poses several challenges especially when researchers or practitioners
implement it within a learning context as a prolonged activity in time. This paper
first outlines the background of game design and participatory game design with
children. Then, it presents the approach used by the authors for participatory game
design with primary school children. The experiences matured from previous field
studies, executed applying the approach, resulted in a set of guidelines useful for
practitioners and researchers for conducting participatory game design with primary
school children.

1 Introduction

This paper presents a participatory design approach to game design with primary
school children. While at school children work in small groups on game design tasks,
with the support of a game design expert through scaffolding dialogues. The approach
relies on cooperative learning for including all, and for promoting democratic collaboration among children. Moreover, the approach employs gamification for sustaining
playful engagement during a game design tasks. Engagement is achieved via gamified probes, such as a progression map, for conveying children a sense of progression,
control and relatedness through game design process.

The paper first outlines a compact review of game design and participatory design
for digital games, focusing on studies with children. Then it presents the approach,
namely GaCoCo, for participatory game design with primary school children. Finally,
the paper presents the lesson learnt from case studies executed applying the GaCoCo
approach. Such lessons are presented in the form of guidelines for conducting a
participatory game design experience with children.

G. Dodero and A. Melonio(✉)
Free University of Bozen-Bolzano, Bolzano, Italy
e-mail: alessandra.melonio@unibz.it

© Springer International Publishing Switzerland 2016
M. Caporuscio et al. (eds.), *mis4TEL*,
Advances in Intelligent Systems and Computing 478,
DOI: 10.1007/978-3-319-40165-2_5

2   Background

2.1   Game Design Process

Children enjoy playing digital games, but for them, designing games is much harder than playing. Designing games, indeed, is a more complex activity than simply playing with them. According to [1], game design is a complex interaction design process made of a complex set of tasks so as to produce a "good" game design product, which transmits information to game developers and allows for the refinement of the game during development and testing.

In the early design of a game the following main tasks are interleaved: (1) the analysis of the goal of the game and a first ideation of the high level concept of the game, (2) game conceptualization, (3) game prototyping. Game designers usually start with the analysis of the game goal and with the conceptualisation of the game idea. This requires thinking of actions for reaching the goal of the game. When a game is structured into levels, game designers have to set the *core mechanics* for progressing across levels, besides the aesthetics for the interface and interaction, including feedback elements. When the game has a *storyline*, designers have to make it consistent with the overall game mechanics and aesthetics [1]. Such tasks produce game design documents, and are complemented with prototypes. In particular, the *high-level concept document* records the key ideas of the game e.g., the game goal and overall setting. The *character document*, instead, records the design of the main character(s) of the game. If the game has more than one level, the *storyline and progression documents* give a general outline of the players experience from the beginning to the end. Another important document is the *core mechanics document*, specifying rules and challenges, which can be specialized per game level. Prototypes are created alongside documents, ranging from low-fidelity paper-based prototypes to high-fidelity interactive ones. All documents and prototypes can be used for evaluating early game design ideas with game design experts with the intended players, following a user/player centred design approach [1]. Such a long process require careful planning for children's sustained engagement in game design, typically spanning several days of work. It also means planning the associated game design tasks for considering children's expression means and skills, so as to incorporate children's ideas into the design process by sustaining their engagement throughout a game design journey that spans several days.

2.2   Participatory Design for Digital Games

In the literature, there are several general theories and guidelines for designers to involve children in a design process, e.g., [8]. In recent years, usage of participatory design (PD) for digital games has been receiving an increasing attention [7, 10]. PD promotes and relies on social and mutual learning, due to the collaborative nature of the co-design process; learning takes place through participation in group settings, and through exchange and sharing of ideas [12]. Different PD methods and techniques consider the involvement of children as designers of interactive design products, with different degrees of participation [11]. In particular, in [10], Moser outlines several techniques that can be applied when children design games. Her approach follows a child centred game design framework, using PD in several case studies, and requires an active user involvement, where researchers and developers give children a more responsible role. Her framework is designed so as to inspire game design researchers or practitioners on how to involve children through a game design process.

3 Gamified Co-design with Cooperative Learning (GaCoCo)

GaCoCo is a PD method for early game design, first proposed in [4] and experienced in [5, 9], which relies on cooperative learning and gamification of learning. In GaCoCo, children usually work in groups of 3-4 members by using cooperative learning strategies, rules and roles, e.g., [14]. Groups are heterogeneous in terms of learning and social skills, so as to promote and take advantage of differences among group members. Gamification is used in GaCoCo to structure the game design process into game design missions. Each mission has its own game design goal and its own game design tasks, referred to as challenges. Missions build one upon the other, alternating challenges for ideation or conceptualization tasks with challenges for prototyping tasks. All children in a group have the possibility to use different skills and to express their ideas, thereby promoting alternative game design solutions. Gamified probes are also used in GaCoCo to tangibly promote children's engagement through the game design process, especially when this is fragmented over extended periods. More specifically, in line with self-determination theory [6], gamified probes, such as progression maps and symbolic contingent rewards, aim at sustaining children's engagement over time by giving them a tangible sense of progression and control over their game design work. Other gamified probes help in making tangible cooperative rules and roles for sharing and relating to each other.

In GaCoCo, children are the main game designers. However, two adult experts work with them, with specific and different roles. One expert is the game design expert. During a design activity, the expert is assisted by teachers if game design takes place at school. The expert follows each group of children and conducts a formative evaluation of their work, asking clarifications and giving rapid feedback about specific design choices, mainly through dialogue. At the end of a game design mission and at the end of the entire game design process, this expert conducts a summative evaluation of children's game design products. Results of such evaluation must be returned to children across game design missions, so as to allow children to further self-reflect on their game design products. The second expert working with children is experienced in child development. This expert acts as observer, and gathers observation data according to the game design process protocol, maintaining a constant dialogue with teachers and the other researcher concerning the class behavior and children's well-being.

3.1 GaCoCo Missions in a Nutshell

The GaCoCo studies, performed in 2014 and 2015, were conducted with two primary-school classes: one of younger learners, 8-9 years old, and one of older learners, 9-10 years old. In total, the study involved 36 learners, 2 teachers, plus the experts designers. The studies gathered both qualitative and quantitative data concerning children's engagement and cooperation in participatory game design process. During the GaCoCo game design activity, each class prototyped a single game per class (class game). In line with the GaCoCo approach, a class was divided into small groups and each group worked on a game level.

The game design activity required five missions at school, and each mission lasted circa two hours. Missions were split into challenges with specific conceptualization and prototyping techniques for releasing game design documents and prototypes. The first mission served as training for children and for creating group identities: groups chose their group name and their logo. In the second mission, each group developed a game storyline arranging group storyline proposals. In the third mission, each group released the high-level concept document with the game idea; then they used it for prototyping a game character to use as the player in their game levels. In the third and fourth missions, each group conceptualized and prototyped the corrisponding game levels, accordingly with the core mechanics documents. Prototypes were paper-based. Finally, in the fifth mission, each group completed the game level with the winning and losing effects and finally presented their game level to the class. After each mission, children were administered a survey aiming at assessing children's emotions [2].

4 Towards Guidelines for Participatory Game Design

Our field studies undertaken in 2014 and 2015 have been organized in line with the GaCoCo approach described in the previous Section, and have been reported in [3, 5]. According to the evidence gathered in the field studies, the following guidelines resulted as a collection of all the components that a researcher or practitioner should consider for conducting participatory game design with primary school children, as in GaCoCo. The guidelines are clustered into five main categories: (1) how to organize a design activity; (2) what to track in a design activity; (3) participants' roles during the design activity; (4) gamified probes; and (5) game design. The clustering results from a thematic analysis conducted by three interaction and game design experts, and and has been revised in light of field studies outcomes.

4.1 How To Organize a Design Activity

The following guidelines focus on how a researcher should organize a design activity with children, in line with GaCoCo, so as to empower children, and to allow collaboration among them.

(DA-1) Brainstorming sessions should be organized in the ideation stage, for eliciting children's ideas

(DA-2) The design activity should be split into progressive missions with progressive challenges in order to empower children

(DA-3) Each mission or challenge, should be organized with its own clear goal, and such goal should be valuable for participants

(DA-4) Each mission should be differentiated as much as possible w.r.t. other missions, for improving children's engagement in time

(DA-5) Each mission should start with a recap of previous work, so as to create links across missions

(DA-6) The first mission should be easy for all, so to create a positive relaxing atmosphere, to promote mutual trust and engagement in the activity, and to allow designers into train children to the design work

(DA-7) The design activity should provide multiple feedback opportunities, from domain experts and from peers, in each mission and across missions: formative feedback during missions and summative feedback across missions.

(DA-8) The design activity should include cooperative learning strategies for pair, group, or for class work within the design protocol

(DA-9) The design activity should include cooperative learning strategies within groups, e.g. subgroup work (pair work), so as to engage more those members, that otherwise would be less engaged or isolated.

4.2 What to Track in a Design Activity

The following guidelines focus on what in GaCoCo studies should be assessed and tracked,

(TA-1) Children's emotions should be tracked in relation to key challenges or missions

(TA-2) Children's work should be assessed during each mission and across missions, so as to track the progression of children's work over time

(TA-3) Children's engagement, and their experience with the material used for designing (e.g., gamified probes), should be tracked during the activity

4.3 *Participants' Roles During the Design Activity*

A GaCoCo activity, besides the main (early) game designers, i.e. children, considers the presence of adults during the activity. In particular in a GaCoCo activity the following participants should be present: (1) an expert designer, skilled in interaction design and game design; (2) an education expert, skilled in child development, and (3) teachers. This section inspects the main roles and tasks of those adults.

Design Expert's Role

(DR-1) Designers should provide formative design feedback during mission, and summative feedback across missions
(DR-2) Designers should provide more detailed feedback in design stages which have been identified as being critical
(DR-3) Designers should train teachers into the design activity prior to its execution with children; this would allow teachers to be involved as active researchers, to manage the activity according to their role, and would allow the expert to assist teachers during critical design missions

Teachers' Role

(TR-1) Teachers, in their role of being experts in the class behavior, should form groups of 3-5 children, heterogeneous in terms of social and learning skill
(TR-2) Teachers should assign cooperative learning roles to group members, and rotate roles across missions, according to children's skills and to missions
(TR-3) Teachers should illustrate the organisation and material to be used according to the protocol for each mission
(TR-4) Teachers should assist the expert designer in communicating with children for scaffolding purposes.

Education Expert's Role

(ER-1) The education expert should be present during the whole activity in class, typically acting as passive observer
(ER-2) The education expert should gather data (e.g., by taking notes) according to the activity goals. In particular, he/she should record data on class behavior and children's interaction with GaCoCo material, as well as children's engagement during the activity
(ER-3) The education expert should maintain a constant dialogue with teachers, on class behavior and on children's well being

(ER-4) The education expert and the teachers should collaborate during group formation, so as to create groups respecting children's relationships and social and learning skills

4.4   Gamified Probes

These guidelines focus on gamified probes to be used in GaCoCo studies. They explain the purposes of the gamified probes and their main functionalities.

(GP-1) Designers should train children in the usage of gamified probes

(GP-2) Gamified probes for children should be affordable for them, each probe must have its own clear functionality, yet it should be open to different usages, even unforeseen by designers

(GP-3) Gamified probes can be enhanced with technology, e.g., micro-electronics, so as to interact with children, to enhance their engagement in the design activity, as well as to track relevant interaction data

(GP-4) Gamified probes should convey a tangible sense of progression across missions and challenges

(GP-5) Gamified probes should tangibly convey and support cooperative learning roles, rules and strategies

(GP-6) Gamified probes, such as as symbolic rewards, should convey a sense of progression and, at the same time, help children feel in control of their design choices; for instance, rewards might consist on material useful for designing, that children may use to customize their design work

(GP-7) A specific gamified probe should allow children to ask for help from the design expert, when they feel in need of him/her, so as to give children the feeling of being socially related to him/her throughout the activity, and not being left on their own

4.5   Game Design

These guidelines concern specific details on how to design games with children using GaCoCo.

(GD-1) Game design training should be given mission by mission, explaining the mission goal and using concrete examples, already familiar for children

(GD-2) A game design activity should start from a good storyline; if game design takes place at school, the story can be related to a school topic

(GD-3) When design is carried on cooperatively at school, each group should be assigned just a single game level, and the class should prototype an entire class game by using a common story as storyline

(GD-4) Each game level should correspond to a specific episode in the game story-line, and children should elaborate on episode before ideating their game level

(GD-5) Game design should use stimulus cards (e.g.,lenses of [13]) for stimulating children's ideation or reflections over specific game elements

(GD-6) The game design document should be divided into small units (e.g., game idea, game mechanics, game aesthetics), each one presented as a simple form to be filled by answering scaffolding questions

(GD-7) Game design forms should use examples from popular video games

(GD-8) The game design material should allow children to easily prototype their game, e.g, using reference frames for the device under consideration; and prototype material that children are familiar with, or that is immediate to use for prototyping

(GD-9) Designers should assign specific game design roles to group members, and such roles should rotate across missions, e.g., one member manages the static setting (e.g., fixed environment or elements), another the dynamic characters

(GD-10) Designers should establish an interaction model for game (e.g., avatar based) and the form factor, posture and input methods (e.g., small games for tablet)

5 Conclusions

This paper has presented a set of guidelines for conducting a participatory digital game design activity with primary school children. Guidelines emerged from the experiences gained through GaCoCo field studies, and should guide researchers and professionals to performing successful early game design with children in learning context. First, the paper presents the necessary background related to the game design process and the participatory design approach used in literature for designing digital game. Then, the paper continues presenting our approach for participatory game design with children, namely GaCoCo. GaCoCo relies on cooperative learning and gamification. Such a design process is undertaken in a learning context; it includes all children and sustains playful engagement over time, such as school. The paper summarizes the result of a series of case studies and past experience. Such a summary is presented as set of guidelines concerning how to organize a participatory game design process, possibly fragmented over time, so as to include and engage all children.

References

1. Adams, E.: Fundamentals of Game Design, 3rd edn. Allyn and Bacon, Pearson (2013)
2. Brondino, M., Dodero, G., Gennari, R., Melonio, A., Pasini, M., Raccanello, D., Torello, S.: Emotions and inclusion in co-design at school: let's measure them! In: Proceedings of Methods and Intelligent Systems for Technology Enhanced Learning (MIS4TEL 2014). Springer (2015)
3. Dodero, G., Gennari, R., Melonio, A., Torello, S.: There is no rose without a thorn: an assessment of a game design expereience for children. In: 11th Edition of CHItaly, the Biannual Conference of the Italian SIGCHI Chapter (2015)
4. Dodero, G., Gennari, R., Melonio, A., Torello, S.: Gamified co-design with cooperative learning. In: CHI 2014 Extended Abstracts on Human Factors in Computing Systems, CHI EA 2014, pp. 707–718. ACM, New York (2014). http://doi.acm.org/10.1145/2559206.2578870
5. Dodero, G., Gennari, R., Melonio, A., Torello, S.: Towards tangible gamified co-design at school: two studies in primary schools. In: Proceedings of the First ACM SIGCHI Annual Symposium on Computer-Human Interaction in Play, CHI PLAY 2014, pp. 77–86. ACM, New York (2014). http://doi.acm.org/10.1145/2658537.2658688
6. Deci, E.L., Koestner, R., Ryan, R.: A meta-analytic review of experiments examining the effects of extrinsic rewards on intrinsic motivation. Psychol Bull. **125**(6), 627–668 (1999). discussion 692–700
7. Khaled, R., Vanden Abeele, V., Van Mechelen, M., Vasalou, A.: Participatory design for serious game design: truth and lies. In: Proceedings of the First ACM SIGCHI Annual Symposium on Computer-Human Interaction in Play, CHI PLAY 2014, pp. 457–460. ACM, New York (2014). http://doi.acm.org/10.1145/2658537.2659018
8. Mazzone, E.: Designing with Children: Reflections on Effective Involvement of Children in the Interaction Design Process. Phd thesis, University of Central Lancashire (2012)
9. Melonio, A.: Game-based co-design of games for learning with children and teachers: research goals and a study. In: CEUR Proceedings of the Doctoral Consortium of CHItaly 2013 (2013)
10. Moser, C.: Child-centered game development (ccgd): developing games with children at school. Personal and Ubiquitous Computing. **17**, 1647–1661 (2013)
11. Read, J., Markopoulos, P.: Child-computer interaction. International Journal of Child-Computer Interaction (2013)
12. Sanders, E.B., Stappers, P.J.: Co-creation and the New Landscapes of Design. CoDesign: International Journal of CoCreation in Design and the Arts **4**(1), 5–18 (2008). http://dx.doi.org/10.1080/15710880701875068
13. Schell, J.: The Art of Game Design: A Book of Lenses. Morgan Kaufmann Publishers Inc., San Francisco (2008)
14. Slavin, R.E.: Student Team Learning: a Practical Guide to Cooperative Learning. National Education Association of the United States, DC (1991)

Part III
Methodologies for Fostering Social Learning

BoloTweet: A Micro-Blogging System for Education

Jorge J. Gomez-Sanz, Álvaro Ortego and Juan Pavón

Abstract Micro-blogging has become a popular activity nowadays thanks to Twitter. However, micro-blogging is more than a commercial product. There exist free software implementations of the same principles. This work introduces Bolotweet, which is built on top of one of these free software platforms. It is oriented towards enabling an effective teacher-student interaction. The tool offers services for the teacher in order to quick evaluate student progress and assess the lessons' apprehended concepts. Students gain a way of sharing conclusions about what has been learnt, and review concepts presented along the lesson. This paper introduces improvements made to this tool and some experience on its use with real students as well as new experimental data.

Keywords Social networks · Micro-blogging · Bolotweet · Professor-student interaction · Continuous evaluation

1 Introduction

New study plans increase the importance of an efficient and frequent teacher-student interaction, promoting in-class participation of students. In Europe, the Bologna process [Croisier and Parveva, 2013] pursues a higher student monitoring, preparing lecture notes, proposing seminars, exercises, and other tools that make the student role a more active one. This new approach affects the teacher work, because of the additional dedication necessary to satisfy this new teaching requirements. In this context, technologies that reduce this effort, while attaining Bologna's goals, would be welcome.

J.J. Gomez-Sanz(✉) · Á. Ortego · J. Pavón
Universidad Complutense de Madrid, Madrid, Spain
e-mail: {jjgomez,alvorteg,jpavon}@fdi.ucm.es

© Springer International Publishing Switzerland 2016
M. Caporuscio et al. (eds.), *mis4TEL*,
Advances in Intelligent Systems and Computing 478,
DOI: 10.1007/978-3-319-40165-2_6

The approach introduced in this paper bases on communities of collaborative learning where students provide content and work with others who have similar experience level. This kind of learning within a team has proven to be an efficient tool for acquiring knowledge [Williams, 2009]. A teaching method based on the student, and its focus on communities of learning, is an environment fitting the web 2.0 technologies [Mason and Rennie, 2008]. Students play an active role as producers rather than consumers, being the teacher the conductor of this process. Such approach is close to social networks, where micro-blogging ones are specially popular. Micro-blogging is the continuous and collaborative production of micro -annotations, which are short texts with 140 characters maximum and that can contain links or hashtags. A hashtag is a text with no special symbols starting with the hash character #. Users share information through these micro-annotations and interconnect them through reply/like/forward actions or sharing hashtags,

Using this communication philosophy, Bolotweet tool [Gomez-sanz et al., 2010] was developed. In Bolotweet, students and teachers share micro-annotations. The annotation activities are evaluated by teachers through a scoring process. Students have a quick way to get feedback from the professor, check which annotations got good scores, and use them as reference. Also, this scoring is used too to create a ranking and gamify the learning process to some extent. This tool has been used in the Universidad Complutense de Madrid in some subjects of the Faculty of Computer Science.

This paper contributes with new functions and experimental data obtained since the original paper [Gomez-sanz et al., 2010]. The new functions include micro-annotation task management, automatic generation of lecture notes, and other improvements in the user interface to increase the ease of use. The data gives clues about the effectiveness of Bolotweet, showing a relevant correlation between scores received through Bolotweet and the scores obtained in the final theoretical exam.

Following, section 2 presents the new features this version of Bolotweet brings and that are the original contribution of this paper. Section 3 presents how the teaching method is applied. Afterwards, section 4 includes criticism as well as experimental data. Section 5 introduces the related work, focusing on micro-blogging social networks for teaching. Finally, section 6 contain the conclusions and future work.

2   Bolotweet Basic Functionality

Bolotweet's development is built over the StatusNet micro-blogging free software platform. This platform permits to create plugins with new functionality. Being a free software platform, the whole system can be hosted within the institution.

There are other alternatives, like Jisko, Sharetronix or JaikuEngine. Nevertheles, Status.Net has become the *Free Software Foundation* chosen platform for social micro-blogging and this guarantees a long term stability and functionality support that others may not have. Over StatusNet, also known as GNU-Social, the following functions have been developed:

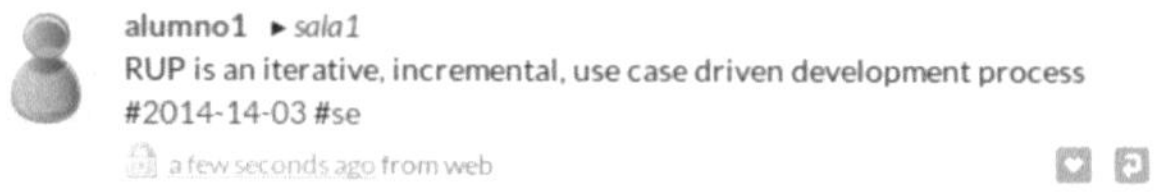

Fig. 1 Micro-annotation example with the hashtags belonging to the lecture day and the an additional lesson tag

Fig. 2 Micro-annotation scoring buttons which are visible only to users playing the role *professor*

– A scoring system for evaluating each micro-annotation of the students. All micro-annotations are subject of receive an evaluation.
– Reporting the scoring status divided by subjects and ordered by the total score achieved by each student. Only the teacher can see all the scores, and students have access to this report in a limited way.
– A new task concept that permits to schedule annotations and let students know they have to produce a micro-annotation with the tags the teacher prearranged.
– Additional "teacher" and "student" roles. This is necessary to adapt the tool to an education environment and assign education-related responsibilities.
– A lecture notes automatic generation procedure. This function generates a summary of the subject progress of all the students, selecting only those better scored ideas.

3 BoloTweet's Teaching Method

Bolotweet teaching method is based on the scoring of micro-annotations. Figure 1 is an example of annotation a student may make. The annotation is presented to the teacher as in figure 2. The teacher chooses a score and hits the button. The result is presented in figure 3. The views offered to teachers and students is different. The student has access to the scores the student obtained, but have none to the scores of others. Teachers have full access to all scores of all students. They even can edit the student scores.

The scored annotations is a quick way for the student to get feedback. And if the annotation had the highest score, it will be collected as an example for the rest of students in the automatically generated lecture notes.

Each subject in Bolotweet is represented as a private group where teachers and students of the same subject are registered. A public group setup has not been tried out yet. Each group has a short nickname which allows students to mention it in

Fig. 3 An scored micro-annotation as seen by the professor. Other students will see a ribbon over this annotation meaning it is a good example for them and the student owning the annotation will not see the edit icon (the icon with a pen)

their annotations, if they need to, without sacrificing too many characters. Every annotation made in the group becomes accessible to all people registered in the group. Students are thoroughly informed that their annotations are published under the creative commons license.

The education tools Bolotweet offers to the teacher are:

- Daily summary. In the last five or ten minutes of the lesson, students have to write down a micro-annotation of an apprehended concept during the lesson. To correctly review later the annotations, they must contain two hashtags, one of with the date in #yyyy-mm-dd format (some students do their annotations the day after), and another written by the teacher and that represents the identifier of the particular daily summary.
- Lesson criticism. It is used by students to share a criticism or suggestions to some part of the lesson, e.g. pointing out the examples given by the teacher were hard to understand and that repeating that part of the lesson would be welcome. This criticism is shared as another micro-annotation with the hashtag **#crit**.
- Question. Students can address questions to other students to answer. This promotes the cooperation among classmates. Such micro-annotations contain the hashtag **#question**

Bolotweet uses mainly the first tool and offers students the second and third ones. Teachers are recommended to structure 50' lessons into a 40' lecturing part and a 10' part dedicated to working with Bolotweet. These last 10 minutes are used to produce a micro-annotation of some relevant idea the student has acquired in the prior 40'. It is expected that this idea is something that has attracted the student's attention the most, but it may refer too to the last part of the lesson the student paid more attention to. During those 10 minutes, the teacher will too evaluate the annotations of the previous day, if he hasn't yet, or the current annotations, as they are produced. Scoring an annotation can be fast, only a matter of pressing a specific score button, or a couple of minutes, if a reply to the annotation of the student is needed to explain what was wrong. The teacher decides which is the best course of action, but, in general, it is the student who, after having a bad score, should inquiry the teacher or just check the support material. The student can reply again to the initial annotation with a corrected version, to prove his interest in the subject and as practical study exercise. In this case, the teacher can set the previous score to 0 and score only the new reply. This is to prevent the student from having extra points for the same annotation. Evaluation process can be assisted by other teachers, if the number of students is too high for a single professor.

Besides the basic micro-blogging function, the platform StatusNet/GNU-Social has additional functions that can be of interest in particular moments. They include *polls*, about any topic the teacher find relevant, and an *event creation service* for announcing special events such as invited speakers attending the lesson.

Nevertheless, the basic support needed by the Bolotweet method had to be developed ad-hoc, over the basic micro-annotation management StatusNet/GNU-Social provides. The new designed functions include the following:

- Micro-annotation task management. Teachers use this tool to create annotation templates student use for a particular exercise the teacher wants to deliver. The micro-annotation template includes the date and a hashtag the teacher configures. The service includes tracking how many students completed the assignment and how many are still pending. The micro-annotation itself incorporates as a new micro-annotation, just like the others, and subject of being evaluated.
- Evaluation of annotations. Each annotation can be scored with a number from 0 to 3. This is a quick feedback sent to students. Scores can be modified anytime by the teacher and the scoring task can be assisted by other teachers. In the later case, scores are computed as the average sum of the scores of all assisting teacher, though other formulas, such as a weighted sum, could be applied.
- Lecture notes. It collects all scored micro-annotations, selects them according to different configurable criteria (such as those most highly scored), and packs them into a pdf report which contains the selected micro-annotations structured per day. Each included micro-annotation include a cite to its author. The later is used to preserve author rights and teach students of the importance of acknowledging authorship. This compilation can be used as lecture notes of collective creation.

4   Evaluation

Bolotweet has been tested in several subjects in the past. This paper includes data from a subject of the 2013-2014 course. There were surveys answered by 31 students. Some of the most relevant results are listed following:

- Frequency of use. The 87.09% of the students used it less than three days a week. None of the surveyed students declared not having used the tool at all. This is consistent with the use only during the lesson.
- Utility. A 22.58% of the students did not find the tool useful, the 38.71% were partially satisfied, and another 38.71% very satisfied.
- Ease of use. This has been one aspect very well welcome, with a 19.35% acknowledging a great ease of use, 74.19% that is was partially easy, and only a 6.45% finding it hard to use.
- Study aid. A 41.93% considers Bolotweet a good support tool as a study aid. It could point out a yet-to-be-improved introduction of the student to the use of tool.

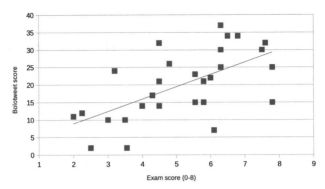

Fig. 4 Scatter chart to show dependency between the score in Bolotweet and the score in the exam

- Importance of scoring micro-annotations. A 58.06% of the surveyed think positively about the obtained feedback from the annotation scores, and a 25.81% think it was of little or none utility. A 16.12% is half sure. This aspect may be improved in future versions through the default incorporation of a text feedback. At this moment, such feedback can already be provided, but it takes more time for the teacher to send the evaluation. Having predefined answers may serve to obtain better results in this area.
- Micro-annotation creation effort. Only a 22.58% of the surveyed consider the micro-annotation requires an important effort, while 41.93% think it is little effort. The remaining 35.48% scores the effort at the middle of the scale. One of the obstacles student point out is the limitation of 140 characters. This survey question is a troublesome one, since, on the one hand, teachers are expecting students to invest effort in order to learn. On the other hand, students are likely to complain when they invest effort. It could be considered that a 22.58% of negative complaints could be a reasonable result, though it is a value to improve too in the future. Ideally, students should feel no effort is invested and that they enjoy learning.

To assess the effectiveness of Bolotweet, the scores obtained in Bolotweet were compared to the scores obtained in the final theoretical exam. This exam had eight questions, so the max score was eight. A scatter chart, see figure 4, shows a regression curve representing the trend of both Bolotweet and the official exam. In fact, the Pearson correlation value is 0.613. Therefore, it can be said that Bolotweet and the exam marks are relatively dependent. It was not possible to state the difference with and without Bolotweet. The experiment was run within the same student group and all students decided to participate.

5   Related Work

The features of micro-blogging social networks, with 140 characters long annotations, and their popularity, have motivated some applied research to education. One of the addressed problems is how to improve the students' motivation. [Grosseck and Holotescu, 2008] have observed that the use of the Twitter commercial platform is a comfortable mechanism for the student. Through this tool, doubts can be answered quickly and easily, they can share opinions, advises, links to external resources, information sources, and others. Altogether, they conform a collaborative knowledge source. Besides, this kind of network promotes the creation of synergies among students, communities, and courses. Compared to [Grosseck and Holotescu, 2008], this work has provided a controlled yet effective way of generating the micro-annotations, see the task concept in section 3. Also, it has addressed the problem of putting together the generated knowledge through the lecture notes mechanism, see section 3.

Some studies [Perez, 2009; Rejon-Guardia et al., 2013] indicate disadvantages too when applied to education. The most remarkable is the need to have someone overseeing the users' activities. Teachers are likely to play this role, and it is not trivial to conduct different active users and make the process a constructive one. On the other hand, the character limit makes difficult to express complex ideas. Typos in the text and the use of slang complicate the understanding of the message. The lack of privacy is a concern too. All annotations can be seen by anyone (private messages are not accounted here). Compared to [Perez, 2009; Rejon-Guardia et al., 2013], with Bolotweet the teacher has a way to deal with typos and bad grammar through scoring. Bad grammar annotations can get low scores. Though annotations are public for all students in the subject, privacy has been taking into account by hiding the scores. Remarkable annotations are highlighted, though, in order to provide with good examples to the students.

6 Conclusions

Bolotweet provides functionality that has not been experimented in other works on social networks for teaching, such as the scoring of students' contributions or the short daily summaries writing.

While running the Bolotweet project, it should be noted students did not misuse the tool or incurred in bad practices, like trolling. Instead, students did their best to express ideas. Cheating, like rewording others' annotations, was uncommon. Due to the low number of students, such situations were easily detected. Larger groups of students may require additional support on behalf Bolotweet, though.

Another aspect of the tool is its capability to give the teacher a snapshot of the class progress in terms of micro-annotations per lesson and/or coverage of the lesson concepts on behalf the students. A low number of micro-annotations may be associated to an excessive complexity of the lesson or to defects in the way the lesson was introduced. Micro-annotations focused on a few concepts from the total set included in the lecture can be a symptom of a lack of attention of the students or a need to improve the teacher's lecturing techniques. The invested effort on behalf the teacher for scoring evaluations is low. Involved teachers confirm the use of Bolotweet is not a big overhead for them.

As criticism, students demand a better way to correct micro-annotations. They realize too late that they had grammar mistakes and typos, and cannot edit once the micro-annotation is published. They are recommended to re-publish the corrected sentence as a reply to the wrong one. Nevertheless, they don't find it satisfactory. The access to the system through mobile devices need to be redesigned, too. Style sheets are used to adapt the HTML interface to different devices, but it still does not work in every screen size.

Acknowledgements This project has been funded by the Teaching Innovation Project (PIMCD 370 - 2014) from the Universidad Complutense de Madrid, collaboration grant 2013-2014 from the Ministry of Education, Culture and Sport of Spain, and the ColoSAAL project (TIN2014-57028-R) by the Ministry of Economy and Competitivity of Spain.

References

Crosier, D., Teodora Parveva, T.: The Bologna Process: Its impact on higher education development in Europe and beyond. Fundamentals of educational planning 97. UNESCO (2013)

Gmez-Sanz, J.J., Cossío, C.G., Fernndez, R.F., Pavón, J.: Microblogging in the european higher education area. In: 1st International Conference on European Transnational Education (ICEUTE 2010), pp. 43–50 (2010)

Williams, S.M.: The Impact of Collaborative, Scaffolded Learning in K-12 Schools: A Meta-Analysis. Metiri Group (2009)

Mason, R., Rennie, F.: E-learning and Social Networking Handbook: Resources for Higher Education, Routledge (2008)

Grosseck, G., Holotescu, C.: Can we use twitter for educational activities? In: Proceedings of the 4th International Scientific Conference eLearning and Software for Education (eLSE 2008) (2008)

Perez, E.: Professors experiment with Twitter as teaching tool. Journal Sentinel (2009)

Rejon-Guardia, F., Sanchez-Fernandez, J., Munoz-Leiva, F.: The acceptance of microblogging in the learning process: the BAM model. Journal of Technology and Science Education (JOTSE) **3**, 31–47 (2013). Science Education (JOTSE), 3 (2013), pp. 31-47

Interaction on Distance Education in Virtual Social Networks: A Case Study with Facebook

Carolina Schmitt Nunes, Cecilia Giuffra Palomino, Marina Keiko Nakayama and Ricardo Azambuja Silveira

Abstract The way as people communicate and interact each other have been changing, especially with the expansion of the internet and the use of social networks. Motivated to a better understanding of this phenomenon in the educational sphere, this research aimed to identify the perception of high school students and tutors about the use of virtual social networks on a distance education course offered in public Brazilian high schools. This work is characterized as a hybrid qualitative study case, classified as pre-dominantly quantitative, but exploratory and descriptive research. The stu-dy case survey data was obtained by questionnaires and the data were collected by an anonymous online survey application filled by students and tutors that participated in the course. The main results confirm what is presented by the most current research on the topic: the use of closed groups in virtual social networks contributes in a significant way for the in-teraction among the participants and for the involvement of the students in the course.

Keywords Facebook · Distance education · Interaction

1 Introduction

The fast and transformative changes that are taking place in the society, especially in the last decades, accelerated by the new communication and information technolo-gies and the popularization of the internet, impact many spheres, including the sphere of education. This impacts result in new communication and interaction scenarios both in classroom teaching as in distance education. In this

C.S. Nunes · C.G. Palomino · M.K. Nakayama · R.A. Silveira(✉)
Universidade Federal de Santa Catarina, Campus Reitor João David Ferreira Lima,
Florianópolis, Santa Catarina, Brazil
e-mail: nunes.carolinas@gmail.com, cecilia.giuffra@posgrad.ufsc.br,
marina@egc.ufsc.br, ricardo.silviera@ufsc.br

© Springer International Publishing Switzerland 2016   61
M. Caporuscio et al. (eds.), *mis4TEL*,
Advances in Intelligent Systems and Computing 478,
DOI: 10.1007/978-3-319-40165-2_7

scenario, comes into de-bate the new improvement possibilities in the teaching and learning process through the use of information and communication technologies. Among them stands out the Facebook Social Network, due to its popularity in the last four years and the signifi-cant amount of users around the world.

The use of Facebook allows people, especially teenagers, to socialize and interact with friends and colleagues from any device with internet access [1]. With the in-creased popularity and having become mature, Facebook is, currently, seen as a pow-erful tool for many areas, including education. In 2011 it was published the document "Facebook for Educators", where the authors show that it is possible to explore the resources of this network for educational purposes [2]. In 2013, Facebook stood out as the ninth best tool for learning, according to the Centre for Learning and Perfor-mance Technologies. Confirming this idea, [3] suggest that the studies about the use of Facebook for educational purposes should continue in order to improve this proce-dure and evaluate if the school reorganization needed to use the Facebook as an educa-tional tool is worth it for the benefits obtained with it.

The motivation of this research emerged from the closing event of the digital train-ing course called Aluno Integrado, in 2013, that is a distance education course spon-sored by the Brazilian Education Department and implemented by the Federal Univer-sity of Santa Catarina in partnership with the Education Secretariat, that was offered for teens in the state school system.

From the testimonies of tutors and students, it was realized the importance of the social network and its potential. On the course edition of 2014, the social network was implemented as a comlementary educational tool for the interaction, using closed groups. From this experience, this work aimed to identify and present the perception, from the point of view of students and tutors, about the use of the Facebook in a distance education course for high school students of the public school system in Santa Catarina. This paper contributes to the advancement of the scientific knowledge by evidencing how the Facebook can help in the decreasing of the dropout in distance learning courses, with teen students of low-income. The delimitation of this research is the digital training course Aluno Integrado, performed in Santa Catarina in 2014.

2   Theoretical Background

In distance education the interaction among participants is considered as a major factor for the learning effectiveness, since the knowledge doesn't exist only individu-ally but also is acquired by means of these interactions [4]. The interactions can be describe and defined in three different types, [5]: student-content interaction, student-student interaction and student-tutor interaction. Accordingly, the use of information and communication technologies that facilitate this interaction is imperative for the good performance of the distance education courses [6]; [7]; [8].

The interactions in distance education are mainly mediated by some information and communication technology and it is observed that with the popularization and expansion of the use of computers and web application tools, the possibilities of interaction – synchronous or asynchronous - were modified, giving quick, safe and efficient techniques to the distance education agents. Following this thought, [9]observes that the distance education mode have more possibilities of communica-tion (synchronous or asynchronous) among the people involved in the process, and that this may result in more effective collective and individual learnings.

[10] pointed out that the use of technologies can improve the motivation and inter-est for learning, contributing to the construction of knowledge. However, they warn that the use of technologies, by itself, is not enough to make difference in the lear-ning and teaching process, and that it is needed to do a reflection, to discuss and propose effective teacher training models for using these technologies.

Among the possibilities that have arisen from the internet, there are the social net-works. Nowadays the Facebook is one of the most well known social network having more than 89 million of users in Brazil. Recent researches about the use of Facebook in education indicate that: (a) this social network is the most popular among the stu-dents [11]; [12] and among the teenagers [13]; (b) the closed groups are tools with a big potential for communication, interaction and collaboration [14]; [15]; [11] and; (c) the use of Facebook can contribute to the students' learning, because it allows the interaction among the users, besides of the collaboration and actively participation of them and also because of the capability to share material and information, allowing the participants to explore their critical thinking [16]; [17].

On the one hand, researches such as [12], which talk about the decision of using the social network, defend the social influence as a very important factor to decide to use this network. Thus, if a student sees a colleague using Facebook, probably, s/he will want to use it too. Following this though, [11] believes that the groups on Face-book, including its interactive characteristics and facility of feedback, have the poten-tial to give to young people what they want and, thus, the Facebook closed groups can provide environments for exchanging of knowledge.

On the other hand, according to some authors the social network has some limita-tions for being used for educational purposes. These authors [15] point out that there is support only for one kind of document, the discussions can not be listed and orde-red and, also, Facebook is not perceived as a safe environment for most users.

It is observed, from the cited references, that the use of Facebook in education is still a field open to new findings. This is explained by the novelty of the social networks and because of the dynamics interactions in them.

This work differs from the previous related works found in regarding to the resear-ched context; The research was conducted with a population of students and tutors who participated in a non-formal and voluntary participation course offered as an ex-tra-curricular activity for high school students in the public school network.

The course was offered totally in a distance education mode. In this context high rates dropout are expected and the Facebook was used as a complementary tool to-ghether with E-Proinfo larning menagenment system as the main learning tool. In this context the main expected role of the Facebook would be help in the iterative communication process to minimize the dropout rate.

3   Methodological Procedure

This research can be characterized as descriptive and empirical. The methodological approach can be classified as predominantly quantitative, the applied methodology was an exploratory and descriptive study case, the study case survey data was obtai-ned by questionnaires. The object of study was the training course Aluno Integrado, edition of 2014 as part of a larger project funded by the Brazilian government to spre-ad knowledge about digital technology in public high schools. The primary data were obtained through anonymous online questionnaires applied to tutors and students of this course. The data was collected from the answers of 158 students and 37 tutors, from September 1st to October 26th of 2014, using exclusively questionnaires that were made available on Google Drive. The participants were users of the social network who were also members of the closed groups created on Facebook for this edition of the course.

The data collection instrument used with the students was adapted from [12] to ga-ther more converging data with the research context within a broader project Aluno Integrado. It was created a questionnaire with 10 multiple choice questions. The first two for knowing the age and genre of the students and the other eight for knowing their habits regarding to the use of internet having as a main focus the Facebook (fre-quency of use, motivation for using it, number of friends, among others).

The questionnaire applied with the tutors was adapted from [18]. This questionnai-re had 17 questions, 10 multiple choice questions and seven open field question. The first four were created for knowing the age, level of education, background and expe-rience in distance education, and the other 13 for knowing the perception of the tutors regarding to the use of the groups in Facebook and the students participation in those groups, besides knowing about the difficulties in the use of the tool and their opinion about the contribution of the network in the students learning. The data analysis was made based on interpretivism. Originally it was aimed to trace a profile of the stu-dents and tutors to, in sequence, identify the perception of both of the groups about the use of Facebook in the course.

3.1   Characterization of Aluno Integrado Course

The Aluno Integrado digital training course in distance mode is a course offered for students of the 1st and 2nd year high school in public schools of the Santa Catarina state in Brazil. The Aluno Integrado is part of the National Program of Continuing Training in Educational Technology called Proinfo and its goa lis to promote the pedagogical use of the Information and Communication Technologies in the public school system. The program brings computers, digital resources and educational con-tents to the schools. In turn, states and counties must warrant the proper structure to maintain the computer rooms and to train the students and teachers for the use of the machines and the technology. The focus is centered in education and technology ai-ming to provide to these students an expansion of knowledge in this field, trying to explore different perspectives in basic education.

The course is offered by mean of the Learning Management System (LMS) E-Proinfo, and the contente and learning activities was developed totally for learning in distance education mode. In this platform the students learn main concepts about this educational modality, about history of computer, hardware (physical part of the com-puter), computer maintenance and operational systems. In this environment the con-tent of the course and the evaluation activities that have to be performed by the stu-dents are available.

In the 2014 edition, 63 classes were created, each one of them had a distance tutor responsible for guiding the students about the content of the course and the activities, and also, for encouraging them to interact in the platform using forums and messages. Besides the tutors there was another role, the tutor guide, in this edition there were three professionals who did this work, each of them had 21 tutors/classes to guide and supervise in regarding to the deadlines, students participation, resistance, evasion, among other.

In 2014 the use of the Facebook social network was expanded (in 2013 it was used the Fan Page and the profile for disclosing the course), by creating closed groups for each of all the virtual classes, including all the tutors, students and project coordina-tors in separated closed groups. At the end, before the classes started, 63 closed group was created by the technical team of the project. The members of each group were the tutor and his/her students of each class, having approximately 50 students.

4   Results

4.1   The Student Perspective

From the answers of the 158 students who participated in the survey, it was possi-ble to trace their technological profile, understand their motivation to use the social network and know how the use of the Facebook can influence in the course progress. In order to characterize the target group, initially is presented the technological profile of these students.

The group of students is predominantly female (59%). The aged of the students is between 14 and 17 years old. The students can access the internet from different pla-ces, from home, from the school and from different mobile devices. This question was created for the participants to have the possibility to select more than one option, the reason for doing this is because the students can have more than one place to ac-cess the internet. For accessing the internet, almost all (95%) use the computer at home, 40% use also at the school, 49% of the students accesses the internet also on the cell phone and 13% use also the tablet (Figure 1).

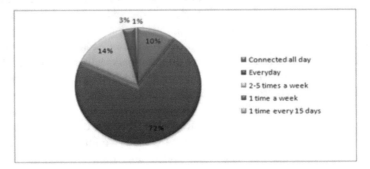

Fig. 1 Frequency of the use of Facebook by students

These results show that the use of cell phones to access the internet begins to pre-vail in comparing to the use of the computer at the school and, even though having just 13% using the tablet, the use of mobile devices (cell phone + tablet) represents a significant amount of the results.

Asked about the motivation to use the Facebook, 89% answered that they use the network to be in touch with their friends, 45% said that wants to be in contact with people that doesn't see a long time ago, 42% try to talk with the colleagues about school tasks and 30% of the students accesses to make new friends. About the number of friends of the students, the majority of them (49%) has a network of between 250 and 1000 friends and 34% has a network of more than 1000 friends.

With these results it was perceived that the main purpose of the students is to make new friends and interact with them. This is reflected in the quantity of friends that they can have, but also, it was observed that a big group of them (42%) use the social network for educational purposes, for being in contact with colleagues and talk about tasks from the school.

Specifically about the use of the Closed Groups tool in the course, the perception of the students is that they were able to be closer to the tutor, using the social network, 94% of the students answered that the communication with the tutor was facilitated with the Facebook. Besides, the students said that this method also helps for the interaction with the colleagues of the class (57%) and that the doubts were answered more quickly by the tutor (47%). Thus, the use of groups increased the interaction between the students and facilitated the communication of them with their tutors, whom were able to help them in less time.

In regard to the frequency of the use of Facebook, some of the students said that they are connected during all day. Also, almost all of them are connected everyday. About the difficulty of the use of the network, 99% said that they don't have problem using the Facebook.

With these results it can be concluded that the insertion of the social networks in education can be performed without causing any frustration to the students for not having much knowledge about the tools offered by the network, because they affirmed that they know and are familiar with the environment, not having difficulties using it.

4.2   The Tutor Perspective

The function of the tutor in the Aluno Integrado course is to guide, monitor and evaluate the student during the course. This process is made answering the questions about the content and the course progress, through the interaction between the tutor and the student using the different communication tools. The communication tools defined in this course were: forum and message in the e-proinfo learning menagen-ment environment, institutional e-mail, phone and Facebook.

The group of tutors participants of the survey is heterogeneous in regards to the age and knowledge and, also, regarding the perception of the use of the social network for educational purposes. The participants are between 24 and 58 years old, the majority (76%) has specialization or master (19%). The education of the tutors is characterized by the heterogeneity: computer science, information systems, languages, administra-tion, accounting, arts, pedagogy, history, mathematics, biology and physical educa-tion. In regard to the experience in distance education, 49% of the tutors have until three years of experience, 21% have between three and five years and 30% have more than five years of experience. However, the most of them (68%) have never used Fa-cebook for educational purposes.

In general the tutors find out that the most of the students use this social network, confirming the researches of [13]. From the tutor's point of view, is easier to be in contact with the student using the Facebook, than using the email or phone, because the students are online everyday. Another perception is that the communication through the network makes the course be more attractive and interesting.

A new aspect is the issue of sense of belongingness. According to the tutor 14 "they view their friends that are together in the class and, being users of the tool, they give 'likes' in the posts. It is favorable to the disclosure of the activities and deadlines of tasks". This issue is raised for [12], confirming the impacts of the social influence in the decision to use or not the network.

The perception of the tutors is not unanimous in regard to the effectiveness of the use of Facebook. There is a group of tutors who believe that the creation and the use of the closed groups for each class played a key role in the exchange of additional content, clarifying questions and in the student engagement, being characterized as a facilitator environment for learning.

For this group of tutors the network still works as an incentive of the communica-tion, making easy the contact with the student in less time. This can be perceived when the tutor 08 said: "I perceived that the contact with the students becomes more frequent, because is a tool that the students like and use". In the same sense, tutor 04 said: "always (and only in this way) they are in contact through Facebook, in everything they need, in any situation, and always I am able to help and solve the problem, in a quick and efficient way, almost immediately, and this is a facilitator and is a motivation in the participation and continuity in the course". Tutor 18 still makes a relation between the participation in the group and the contributions in the forum of the course: "once they establish the communication through Facebook, they feel more comfortable to interact, also, through the forum".

In contrast, a group of around 16% of the tutors believe that the use of this tool doesn't have relevant impact in the learning of the students. Tutor 28 said that: "what makes difference for the student to learn or not is her/his engagement with the course, the responsibility level and the time that s/he takes to do the course". In this direc-tion, tutor 37 affirms that the most of the students don't understand that "Facebook can also be a learning tool and an exchange information tool, and not just entertain-ment".

Based on the reports is clearer that the use of the tool alone doesn't warrant the im-provement in the teaching learning process, as have been pointed for [10], requiring both awareness and maturity of the students as management with planning, control and monitoring of network usage.

An additional issue that has emerged from the data is the use of another communi-cation tool, Whatsapp. According to tutor 15 "the Whatsapp is being much more used than Facebook. The Facebook group is more for the students that are not in the What-sapp group, because who are in this group prefers to communicate through it". This testimony shows the trend of popularity of this new communication tool among the teenagers.

The results encountered in this research, contextualized in the scenario of the public schools of the state of Santa Catarina, confirm the main points shown by the interna-tional literature and emphasize the point of view of the students and tutors, main actors in distance education courses.

5   Final Remarks

The main contribution of this research is to demonstrate that the use of groups in Facebook as interaction spaces among the students and tutors is effective and brings benefits for the engagement of the students in the studied course. It was possible to notice an increase in the interaction levels among the students, among students and tutors and among students and course content. This fact affected directly the decrease of the course dropout. In this research the groups in

Facebook were important tools in the communication and interaction with the students of the course, allowing the tutor to be closer of them, strengthening the relations between all the participants of the course.

As we pointed above, it can be concluded that the insertion of the social networks in education can be performed without causing any frustration to the students for not having much knowledge about the tools offered by the network. But the group of tutors participants can be heterogeneous in regards to the age and knowledge and, also, regarding the perception of the use of the social network for educational purpo-ses.

We observed that the tutors find out that the most of the students use this social network, and it is easier to be in contact with the student using the Facebook, than other tools and the communication through the network makes the course be more attractive and interesting.

It is important to highlight that the use of these groups is possible, because of the previous knowledge of the students about the Facebook and their preference to be online, with friends or to interact in the platform. In contrast, it was evident that the existence of the groups in Facebook doesn't ensure its effective use, being needed planning and pedagogical guide for having a successful tool for education.

This knowledge can help in the creation and implementation of communication and interaction strategies through the social networks in distance education courses focused, specially, for teenagers. Contributing for the improvement of the quality of the course, and for the increase of the interaction of the students with their colleagues, and with the tutor, decreasing the distance existent in a distance education course.

As a future research works, we intend to perform additional studies to identify if the participation and interaction of the student in the group of Facebook has any in-fluence, positive or negative, in her/his academic performance in it. Additionally we consider the integration of a regression analysis to provide a more clear and scientific framework using a multinomial logistic regression considering each technological profile of students and tutors. By this improvement in this research, participants would be classified according to their assiduity: frequency on platform (for students) and number of assignments (for tutors), creating a sort of index.

References

1. Cheung, C.M.K., Chiu, P.Y., Lee, M.K.O.: Online social networks: Why do students use Facebook? Computers in Human Behavior (2010)
2. Phillips, L.F., Baird, D., Fogg, B.J., Facebook for Educators: Disponível em: https://www.facebook.com/safety/attachment/FacebookforEducators.pdf.   2011. Acesso em 3. set. 2014
3. Muñoz, C.L., Terri, T.: Opening Facebook: How to Use Facebook in the College Class-room. First Monday (2011)

4. Chau, C-Y., Hwu, S-L., Chang,C-C.: Supporting interaction among participants of on-line learning using the knowledge sharing concept. TOJET: The Turkish Online. Jour-nal of Educational Technology **10**(4) (2011)
5. Moore, M.G.: Three types of interaction. The American Journal of Distance Education **3**(2) (1989)
6. Moore, M., Kearsley, G.: Educação a distância: uma visão integrada. Thomson, São Paulo (2007)
7. Anderson, T.. Modes of interaction in distance education: recent developments and research questions. In: Moore, M., Anderson, G. (eds.) Handbook of Distance Education, Erlbaum, NJ (2003)
8. Hedberg, J., Sims, R.: Speculations on design team interactions. Journal of Interactive Learning Research **12**(2), 193–208 (2001)
9. Moran, J.M.: A educação que desejamos: Novos desafios e como chegar lá. Papirus, Campinas (2007)
10. Feldkercher, N., Manara, A.S.: O uso das tecnologias na educação à distância pelo professor tutor. RIED. Revista Iberoamericana de Educación a Distancia **15**(2), 31–52 (2012)
11. Yunus, M. Md., Salehi, H.: The Effectiveness of Facebook Groups on Teaching and Im-proving Writing: Students' Perceptions. International Journal of Education and Information Technologies **6**(1) (2012)
12. Artega, S.R., Cortijo, V., Javed, U.: Students' perceptions of Facebook for academic purposes. Computers and Education (2014)
13. Lenhart, A., Madden, M.: Teens, privacy, & online social networks. Pew Internet and American Life Project Report (2007)
14. Schroeder, J., Greenbowe, T.: The chemistry of Facebook: using social networking to create an online community for the organic chemistry laboratory. Innovate: Journal of Online Education **5**(4) (2009)
15. Wang, Q., Woo, H.L., Quek, C.L., Yang, Y., Liu, M.: Using the Facebook group as a learning management system: An exploratory study. British Journal of Educational Technology **43**, 428–438 (2012)
16. Selwyn, N.: Faceworking: exploring students' education-related use of Facebook. Learning, Media and Technology **34**(2), 157–174 (2009)
17. Tapscott, D., Williams, A.: Innovating the 21st century university: it's time. Educause Review **45**(1), 17–29 (2010)
18. Pi, S.-M., Chou, C.-H., Liao, H.-L.: A study of Facebook Groups members' knowledge sharing. Computers in Human Behavior (2013)

Part IV
Technology-Based Learning Experiences

A "light" Application of Blended Extreme Apprenticeship in Teaching Programming to Students of Mathematics

Ugo Solitro, Margherita Zorzi, Margherita Pasini and Margherita Brondino

Abstract In this paper, we analyse an application of the eXtreme Apprenticeship (XA) methodology, in a blended form, with a reduced set of human and software resources. The study was conducted at the University of Verona, in the context of the course "Programming with Laboratory" with 170 participants enrolled at the first degree in Applied Mathematics, throughout three different academic years. We analyse the earliest period of lessons, when the fundamentals of programming are introduced. During the first two years, students were trained with a traditional teaching method; the last group was trained using the XA teaching model. The outcomes showed a tangible improvement of learning outcomes in students trained with XA compared with the traditional teaching method. Possible refinements of XA method in our case study and in other educational contexts are discussed.

1 Introduction

The ability to analyse problems, design a solution and finally fulfil it using the available resources (in short, *programming*) requires a set of knowledges and skills related to the so-called *computational thinking*, whose importance has been largely recognized [8, 15]. Teaching programming has proved to be a critical task and becomes particularly interesting and challenging when this matter is proposed to students of non-vocational curricula. In Italy, informatics is a marginal discipline in most non-technical high schools. As a consequence, many college students of scientific disciplines (different from computer science), have little knowledge in informatics. Freshmen typically address basic programming courses as beginners, without previously acquired skills. Notwithstanding, computational thinking attitude

U. Solitro(✉) · M. Zorzi
Department of Computer Science, University of Verona, Verona, Italy
e-mail: {ugo.solitro,margherita.zorzi}@univr.it

M. Pasini · M. Brondino
Department of Human Sciences, University of Verona, Verona, Italy
e-mail: {margherita.pasini,margherita.brondino}@univr.it

© Springer International Publishing Switzerland 2016
M. Caporuscio et al. (eds.), *mis4TEL*,
Advances in Intelligent Systems and Computing 478,
DOI: 10.1007/978-3-319-40165-2_8

is nowadays essential and some basic masteries in this field are unavoidable, in particular for technical and scientific subjects. In mathematics curricula the computer (more precisely, the tools based on it) is a key device for students. The general shortage of primary informatics skills (algorithmic thinking, basics coding abilities, experience in problem-solving activities) and the heterogeneous classes' composition, determine a number of initial troubles, such as misconception in the basics of informatics, difficulties in conceiving algorithmic solutions and designing the corresponding code, understanding mistakes, and more. Many students complete the first course in Informatics with a sufficient general knowledge of the topics, but without satisfactory practical competencies. For this reason, we decide to improve our methodology in teaching programming by adopting some techniques inspired by the *eXtreme Apprenticeship* (XA) paradigm [13]. We choose the following case study: the course "Computer Programming with Laboratory", offered to first-year students of the Bachelor Degree in Applied Mathematics at the University of Verona. We follow the main ideas of eXtreme Apprenticeship, and define a "light" version of the paradigm by adapting the methodology to the particular educational context we are considering (see Section 2.2). We employ the available *E-learning* platform elearning.univr.it, based on *Moodle* (moodle.org), as a technological tool in support of the teaching-learning process. We analyse our ongoing experience, and we measure the effectiveness of the XA teaching model also when it is applied with a reduced set of resources.

1.1   The eXtreme Apprenticeship Method

A promising perspective in promoting computational thinking is the Cognitive Apprenticeship (CA) learning model [3], a method inspired by the apprentice-expert model in which skills are learned within a community, through direct experience and practice guided by an expert of the subject. The main idea of Cognitive Apprenticeship is to focus on the teaching/learning process rather than just on the final "product". Cognitive apprenticeship is based on three separate stages: *modeling*, *scaffolding* and *fading*. In the modeling stage, the teacher gives students a conceptual model of the process. Lessons are principally based on presenting work examples in an interacting and active manner: the teacher explains the decisions made during the process step by step. During scaffolding stage students solve exercise under the guidance of an experienced instructor. Students receive hints to be able to discover the answers to their questions themselves. The fading stage of apprenticeship learning is reached when students are able to master tasks by themselves.

Cognitive Apprenticeship has had many applications in teaching programming. *eXtreme Apprenticeship* (XA) is an extension of the Cognitive Apprenticeship model, which emphasizes communication between teacher and learners during the problem-solving process. This approach promotes learners' intrinsic motivation, which is a positive antecedent of performance. This methodology has been developed and being actively practised since 2010 at the University of Helsinki (Finland) [13, 14] for teaching programming. In XA hours devoted to frontal lessons are drastically reduced, and students are encouraged to solve problems themselves in a guided manner and in a non-interfering environment. Exercises are the most important aspect of the learning process: student's apprenticeship starts immediately, gradually, and it is continuously monitored, by assigning programming exercises. The difficulty of exercises has to be slowly incremental: as pointed out in [13], each new exercise has to master a minimum amount of new material on top of previous exercises. In this way, students acquire new skills facing with a measurable amount of work to be done. Programming exercises have also a positive impact on the motivational side: there is no need for any external motivating factors, since success in learning itself promotes further motivation (ee, for example, [1, 2, 10]). With the application of XA method, the Finnish research team RAGE has obtained excellent results in programming introductory courses [13]. The method has also been successfully applied in a *blended* way (i.e. with online support) in Bolzano (Italy). In [4, 5, 6, 7] are described results obtained in teaching operating systems and informative systems.

XA is currently applied in two programming course in Verona, offered at the bachelor degree in Computer Science and at the bachelor degree in Applied Mathematics respectively. As previously said, the latter represents our case study.

2   Method

2.1   *Participants*

The study was conducted at the University of Verona, with 170 participants (50% males, mean age: 19) enrolled at the first degree in Applied Mathematics; the course is Computer Programming with Laboratory, throughout three different academic years (2013/14; 2014/15; 2015/16). We analyse the first period (8 weeks) of lessons, when the fundamentals of programming are introduced.

During the previous two years, students were trained with a traditional teaching method (TT). The current group is trained using the XA teaching model; some students of the last group actually didn't participate in the training and delivered no one of the expected programming tasks connected with the XA experience, and for this reason were excluded from the XA group. The final sample consists of 114 students (50% males) for the TT condition and 48 students (54% males) for the XA condition.

2.2   *Procedure*

The first two months are spent to an introduction to algorithms and programming, partly using the programming language `Python`. The teaching is organized in two kinds of activities.

– a *theoretical part*: a wide introduction to programming in a lecture hall;
– a *practical part*: programming experience in the computer laboratory.

In the TT way (A.A. 2013/14 and 2014/15) the two activities are strongly distinct and have different focuses. In the lecture hall the teacher introduces the general concepts and some exercises. In the laboratory the full or partial solutions of few exercises are presented; the students are encouraged to complete some exercises. Additional activities are proposed as a homework. Teacher and assistants (graduate or Ph.D. students) give limited practical support during the laboratory sessions.

In the XA way the teacher, during the non-practical lessons, usually introduces the tools necessary to understand the practical activities and some general concepts about informatics and programming. When possible, some exercises are proposed. Students have to attend introductory lessons and take part to exercise sessions. During home works they can receive feedback and support in the laboratory or through the E-learning platform from the teacher or the assistants. Every two weeks students are asked to submit the result of an activity that will be evaluated.

The rigorous application of XA method requires a lot of human resources [11] and, if possible, semi-automated correction tools (see, for instance, `Test My Code` [12]). In the "Verona setting" we had to overcome some difficulties of our educational context: few hours of practical lessons with respect to XA standard, limited support by laboratory assistants. So we have adapted the teaching routine and, in a sense, our application of XA method can be defined *light*: we could not fully apply the paradigm but, notwithstanding, we followed its main guidelines.

We claim and prove (see Section 3) that XA also works in this particular situation, which is otherwise realistic: a typical educational setting has to face a limited amount of resources and, as a consequence, a flexible adjustment of the teaching method may be mandatory.

2.3   Research Design and Data Analysis

At the end of the first period students are encouraged to take a partial exam (*test*) considered as part of the final examination. The following parameters are considered for the evaluation: correctness of the solution, logical structure and good programming practices. The test it the only learning outcome available for all three academic years token into account in this study, and for this reason was considered to verify the efficacy of XA compared with the traditional teaching model. The test consists of two parts: a general theoretical section, in which the knowledge of fundamental notions (e.g. the definitions of compiler, interpreter, specification...) are verified; a practical, programming section, where students must solve a few exercises of increasing difficulty about programming competences and problem solving skills.

The evaluation of the test produced a quantitative score, which was normalized in the range 0-1 to allow the comparison among the three different academic year and the comparison between the theoretical and the programming outcomes.

At the end, three different-even if related-quantitative dependent variables were considered: total score (TOT), theoretical score (TH), and programming score (PR). Two different data analyses were carried on.

1. Considering the whole sample, a quasi-experimental design was used, with the teaching method as the independent variable, with two conditions (XA vs TT), and the three learning outcomes as the dependent variables.
2. Considering the XA group as a sub-sample, a quasi-experimental design was used, with total number of delivered programming tasks as an independent 3 level variable (0/1; 2; 3), and the same learning outcomes.

In a third descriptive analysis, the quantitative score was also categorized in 6 different judgements A, B, C, D, E, F where 'A' represents the highest level and 'F' the lowest one.

3 Results

A first set of analyses considered the entire sample. Figure 1 shows the average exam score considering the three learning outcomes (TOT, TH and PR) for the two groups (XA and TT). Three separate two-way ANOVAs were run, with the teaching method as a two-level between-group factor, also considering the two-level factor "sex", in order to control for a possible sex effect. First of all, no sex effect was found, nor interaction sex by teaching method for the learning outcome TOT nor for the two separate learning outcomes TH and PR. The main effect of the teaching method was found for TOT ($F(1,158)=17,23$; $p<.001$) even if with a small effect size ($\eta^2 = .10$), and also for TH ($F(1,158)=29.55$; $p<.001$), with a medium effect size ($\eta^2 = .16$), and for PR ($F(1,158)=4,95$; $p<.05$), with a small effect size ($\eta^2 = .03$). In all cases, XA group showed the best results (dependent variable "TOT", XA: mean=.73, SD=.03; TT: mean=.59, SD=.03; Dependent variable "TH", XA: mean=.79, SD=.03; TT: mean=.62, SD=.02; Dependent variable "PR", XA: mean=.61, SD=.03; TT: mean=.52, SD=.02).

These results seem in line with the expected better outcomes due to the XA teaching model, even if the last result with a small effect size on programming task, needs to be more deeply explored.

In order to verify whether the performance, in the XA group, is connected with the persistence in doing the assigned tasks, another ANOVA was performed, in the sub-sample of the XA group, using the three-level variable with the total number of delivered programming tasks (0/1; 2; 3) as the between-group factor and the three exam scores as the dependent variable. As expected, the higher the number of delivered programming tasks, the better the performance (see Figure 2). This effect is significant for TOT score, due to the effect on TH and not on PR (TOT: $F(2,53)=3,98$, $p<.05$; TH: $F(2,53)=4,75$, $p<.05$).

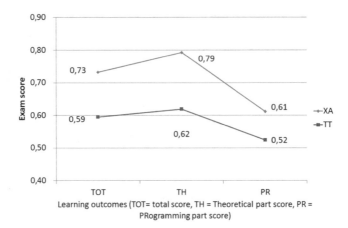

Fig. 1 Average exam scores in the two groups - Traditional Teaching method group (TT) and XA group - considering the overall score (TOT), and the score in the two parts: Theoretical (TH) and Programming (PR) part.

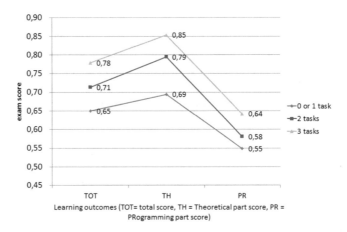

Fig. 2 Average exam scores in the three XA-level groups considering the overall score (TOT), and the score in the two parts: Theoretical (TH) and Programming part (PR).

The last analysis concerns the judgement in the exam, in 6 levels, from A to F. Figure 3 shows that judgements A and B are more frequent for XA students, whereas judgements E and F are more frequent for TT students. Percentages of central categories C and D are similar in the two groups.

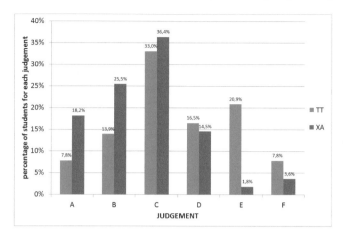

Fig. 3 Percentage of students for each judgement (from A=excellent to F=poor), separately for TT group and XA group

4 Conclusions

The aim of the present study is to explore the potential advantages of the XA teaching model on the learning outcomes in teaching programming to university students.

Our first results are encouraging: the XA method drastically reduces the number of low results. A different teaching perspective to programming seems to have a positive effect both in the "theoretical" part and in the "practical" one. The improvement of the performance is less evident in more advanced exercises that in general require more experience and, a posteriori, appear to be slightly more difficult for the current year.

The combination of a more practical teaching approach, an active support, a constant stimulation to cope with the difficulties and the opportunity to receive feedback in the lab sessions and through the e-learning platform, have contributed to a more effective participation of the students in the activities.

We are planning to extend the XA method to the teaching of programming in different curricula and also explore new potential applications. A wider application of the methodology will require the solution of some problems in connection with the shortage of human resources and the necessity of a significant revision of the e-learning services.

In the next years, we hope to collect more data from different XA trained classes, and obtain fruitful hints in order to refine XA methodology in our case study and in other contexts. Database courses addressed to students of liberal arts could be, for instance, an intriguing context to investigate.

Finally, we think that eXtreme Apprenticeship offers benefits both for students and teachers. During the lessons period, an XA teacher collects a huge amount of delivered programming exercises. This offers an unprecedented overview about errors students commit in solving programming tasks. A careful analysis of students' failures will provide a precious feedback about learning cognitive processes and teaching methodology.

We think that this investigation, together with an analysis of students' motivations and emotions, will provide useful tools for overcoming the beginners' difficulties in learning programming.

References

1. Bergin, S., Reilly, R.: The influence of motivation and comfort-level on learning to program. In: Proceedings of the 17th Workshop on Psychology of Programming Interest Group (PPIG 2005), pp. 293–304 (2005)
2. Brondino, M., Dodero, G., Gennari, R., Melonio, A., Pasini, M., Raccanello, D., Torello, S.: Emotions and inclusion in co-design at school: Let's measure them. In: Methodologies and Intelligent Systems for Technology Enhanced Learning, pp. 1–8. Springer (2015)
3. Collins, A., Brown, J., Holum, A.: Cognitive apprenticeship: Making thinking visible. American Educator **6**, 38–46 (1991)
4. Del Fatto, V., Dodero, G.: Experiencing a new method in teaching Databases using Blended eXtreme Apprenticeship. In: Proceedings of 21st International Conference on Distributed Multimedia Systems (DMS 2015) (2015)
5. Del Fatto, V., Dodero, G., Gennari, R.: Operating Systems with Blended Extreme Apprenticeship: What Are Students' Perceptions? IxD&A **23**, 24–37 (2014)
6. Del Fatto, V., Dodero, G., Gennari, R.: Assessing student perception of extreme apprenticeship for operating systems. In: Proceedings of14th International Conference on Advanced Learning Technologies (ICALT), pp. 459–460 (2014)
7. Dodero, G., Di Cerbo, F.: Extreme apprenticeship goes blended: An experience. In: Proceedings of 12th International Conference on Advanced Learning Technologies (ICALT), pp. 324–326 (2012)
8. Gander, W. et al.: Informatics education: Europe cannot afford to miss the ACM 2013. http://europe.acm.org/iereport/ie.html
9. Hautala, T., Romu, T., Rämö, J., Vikberg, T.: Extreme apprenticeship method in teaching university-level mathematics. In: Proceedings of 12th International Congress on Mathematical Education Program Name, COEX, Seoul, Korea, July 8-15 (2012)
10. Jenkins, T.: The motivation of students of programming. In: Proceedings of the 6th Annual Conference on Innovation and Technology in Computer Science Education, Canterbury (UK), pp. 53–56 (2001)
11. Kurhila, J., Vihavainen, A.: Management, structures and tools to scale up personal advising in large programming courses. In: Proceedings of the 2011 Conference on Information Technology Education, pp. 3–8. ACM (2011)
12. Pärtel, M., Luukkainen, M., Vihavainen, A., Vikberg, T.: Test My Code. International Journal of Technology Enhanced Learning 2 **5**(3–4), 271–283 (2013)
13. Vihavainen, A., Paksula, M., Luukkainen, M.: Extreme apprenticeship method in teaching programming for beginners. In: Proceedings of the 42nd ACM Technical Symposium on Computer Science Education (SIGCSE 2011) pp. 93–98 (2011)
14. Vihavainen, A., Luukkainen, M.: Results from a three-year transition to the extreme apprenticeship method. In: Proceedings of IEEE 13th International Conference on Advanced Learning Technologies (ICALT), pp. 336–340 (2013)
15. Wing, J.: Computational Thinking. Communications of the ACM - Self managed systems **49**(3), 33–35 (2006)

A Soft Computing Approach to Quality Evaluation of General Chemistry Learning in Higher Education

Margarida Figueiredo, José Neves and Henrique Vicente

Abstract In contemporary societies higher education must shape individuals able to solve problems in a workable and simpler manner and, therefore, a multidisciplinary view of the problems, with insights in disciplines like psychology, mathematics or computer science becomes mandatory. Undeniably, the great challenge for teachers is to provide a comprehensive training in General Chemistry with high standards of quality, and aiming not only at the promotion of the student's academic success, but also at the understanding of the competences/skills required to their future doings. Thus, this work will be focused on the development of an intelligent system to assess the Quality-of-General-Chemistry-Learning, based on factors related with subject, teachers and students.

Keywords General Chemistry · Higher Education · Logic Programming · Knowledge representation and reasoning · Artificial Neural Networks

1 Introduction

In recent decades technological courses in different areas become increasingly essential in modern societies. Many of these areas are related with basic sciences

M. Figueiredo
Departamento de Química, Centro de Investigação em Educação e Psicologia,
Escola de Ciências e Tecnologia, Universidade de Évora, Évora, Portugal
e-mail: mtf@uevora.pt

J. Neves(✉) · H. Vicente
Centro Algoritmi, Universidade do Minho, Braga, Portugal
e-mail: jneves@di.uminho.pt, hvicente@uevora.pt

H. Vicente
Departamento de Química, Escola de Ciências e Tecnologia,
Universidade de Évora, Évora, Portugal

M. Caporuscio et al. (eds.), *mis4TEL*,
Advances in Intelligent Systems and Computing 478,
DOI: 10.1007/978-3-319-40165-2_9

like chemistry, physics or biology, making it necessary to include such curricula in the study plans of these courses. Indeed, only a solid background in these areas will give students a multidisciplinary vision of the problems. However, frequently these disciplines are not properly framed in the curricula or adjusted to the student's previous knowledge. In particular, in General Chemistry (GC), some studies show that it is seen as a difficult and boring discipline, and thus jeopardize the role that it should play in the student's training [1, 2]. Therefore, the main challenge is to highlight the relationships between academic syllabus and daily life, aiming to avoid the indifference of some students when attending GC. Therefore, teachers must create stimulant-learning environments that may awake students' interest [2]. However, success in studies is a complex phenomenon that involves a large number of factors, some of which depend on the student, others on the teachers, and even on the institutions [3]. Thus, the assessment of university education should be focused not only on the student's progress but also on the development of abilities and skills necessary for access to employment, lifelong education and personal success [4]. Consequently, it is difficult to assess the Quality Evaluation of General Chemistry Learning (QEGQL) since it needs to consider different conditions with complex relations among them, where the available data may be incomplete/unknown (e.g., absence of answers to some questions presented in the questionnaire), and/or contradictory (e.g., questions relating to the same issue with incongruous answers). In order to overcome these difficulties, the present work reports the founding of an intelligent computational framework that uses knowledge representation and reasoning techniques to set the structure of the information and the associate inference mechanisms, i.e., it will be centered on a Proof Theoretical approach to Logic Programming (LP) [5], complemented with a computational framework based on Artificial Neural Networks (ANNs), selected due to their dynamics like adaptability, robustness, and flexibility [6].

2   Knowledge Representation and Reasoning

Knowledge and belief are generally incomplete, contradictory, or even error sensitive, being desirable to use formal tools to deal with the problems that arise from the use of partial, contradictory, ambiguous, imperfect, nebulous, or missing information [5, 6]. The LP paradigm has been used in knowledge representation and reasoning in different areas, such as Model Theory [7], and Proof Theory [5, 6]. In this work the proof theoretical approach is followed in terms of an extension to LP. An Extended Logic Program may be seen as a finite set of clauses given in the form:

$$\{ \ p \leftarrow p_1, \cdots, p_n, not \ q_1, \cdots, not \ q_m$$

$$? \ (p_1, \cdots, p_n, not \ q_1, \cdots, not \ q_m) \ (n, m \geq 0)$$

$$exception_{p_1} \quad \cdots \quad exception_{p_j} \ (0 \leq j \leq k), \ being \ k \ an \ integer$$

$$\} :: scoring_{value}$$

where "?" is a domain atom denoting falsity, the p_i, q_j, and p are classical ground literals, i.e., either positive atoms or atoms preceded by the classical negation sign ← [5]. Under this formalism, every program is associated with a set of abducibles [7], given here in the form of exceptions to the extensions of the predicates that make the program. The term $scoring_{value}$ stands for the relative weight of the extension of a specific predicate with respect to the extensions of the peer ones that make the overall program.

In order to evaluate the knowledge that can be associated to a logic program, an assessment of the Quality-of-Information (QoI), given by a truth-value in the interval $0...1$, that stems from the extensions of the predicates that make a program, inclusive in dynamic environments, is set [8]. Thus, $QoI_i = 1$ when the information is known (positive) or false (negative) and $QoI_i = 0$ if the information is unknown. Finally for situations where the extension of $predicate_i$ is unknown but can be taken from a set of terms, $QoI_i \in]0...1[$. Thus, for those situations, the QoI is given by:

$$QoI_i = {}^1/_{Card} \qquad (1)$$

where Card denotes the cardinality of the abducible or exception set for i, if the abducible or exception set is disjoint. If the abducible or exception set is not disjoint, the clause's set is given by $C_1^{Card} + \cdots + C_{Card}^{Card}$, under which the QoI evaluation takes the form:

$$QoI_{i_{1 \leq i \leq Card}} = {}^1/_{C_1^{Card}}, \cdots, {}^1/_{C_{Card}^{Card}} \qquad (2)$$

where C_{Card}^{Card} is a card-combination subset, with Card elements. The objective is to build a quantification process of QoI and measure one's Degree of Confidence (DoC) on the argument values or attributes of the terms that make a predicate's extension, taking into consideration their domains [9]. Thus, the universe of discourse is engendered according to the information presented in the extensions of such predicates, according to productions of the type:

$$predicate_i - \bigcup_{1 \leq j \leq m} clause_j \left((QoI_{x_1}, DoC_{x_1}), \cdots, (QoI_{x_n}, DoC_{x_n}) \right) :: QoI_i :: DoC_i \qquad (3)$$

where $\bigcup$ and m stand, respectively, for set union and the cardinality of the extension of $predicate_i$.

3   Methods

Aiming to develop a predictive model to assess the QEGQL a questionnaire was designed specifically for this study and applied to a cohort of 127 General Chemistry students. This section describes briefly the data collection tool and how the information is pre-processed.

3.1 Questionnaire

The questions included in the questionnaire were organized into three sections, where the former one includes the questions related with student attendance in different types of GC classes and the number of study hours (Table *Student Related Factors* in Fig. 2). The second one comprises the questions related with the student's opinions about the subject GC (Table *Subject Related Factors* in Fig. 2). The last one includes questions related with the opinion of students about the GC trainer (Table *Teacher Related Factors* in Fig. 2).

3.2 Data Pre-processing

Aiming at the quantification of the qualitative information obtained via the questionnaire, and in order to make easy the understanding of the process, it was decided to put it in a graphical form. Taking as an example a set of 3 (three) questions regarding a particular subject (where the possible answers are *low*, *moderate*, *high* and *very high*) a unitary radius circle split into 3 (three) slices is itemized (Fig. 1). The marks in the axis correspond to each of the possible answers. If the answer to question 1 is *high* the area correspondent is $\pi \times 0.75^2/3$, i.e., $0.19\,\pi$ (Fig. 1(a)). Assuming that in the question 2 are marked the answers *high* and *very high*, the correspondent area ranges in the interval $\pi \times 0.75^2/3 \,\cdots\, \pi \times 1^2/3$, i.e., $0.19\,\pi \,\cdots\, 0.33\pi$ (Fig. 1(b)). Finally, in question 3 if no answer is ticked, all the hypotheses should be considered and the area varies in the interval $\pi \times 0.25^2/3 \,\cdots\, \pi \times 1^2/3$, i.e., $0.08\,\pi \,\cdots\, 0.33\pi$ (Fig. 1(c)). The total area is the sum of the partial ones and is set in the interval $0.46\,\pi \,\cdots\, 0.85\pi$ (Fig. 1(d)). The normalized area is the ratio between the area of the figure and the area of the unitary radius circle. Thus, the quantitative value regarding the subject in analysis is set to the interval $0.46 \,\cdots\, 0.85$.

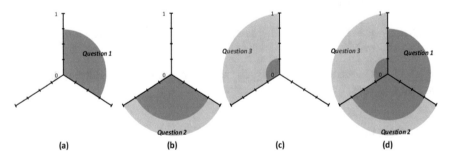

Fig. 1 A view of the questions qualitative evaluation process.

4   Results and Discussion

It is now possible to build up a knowledge database given in terms of the extensions of the relations (or tables) depicted in Fig. 2, which denote a situation where one has to manage information in order to evaluate the Quality-of-Learning of the GC students. Indeed, the *Subject*, *Student*, and *Teacher Related Factors* tables are populated with the responses to the issues presented in the questionnaire, where some incomplete, default and/or unknown data is present. For instance, in the former case the response to the question related with the *coherence syllabus/objectives* is unknown (depicted by the symbol ⊥), while the response to the question associated to the *coherence teaching methodologies/objectives* is not conclusive (*High/Moderate*).

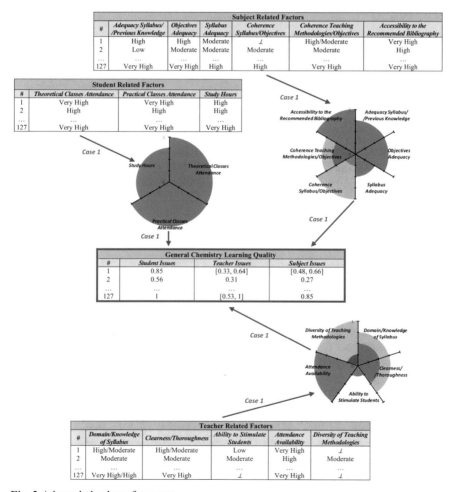

Fig. 2 A knowledge base fragment.

In order to quantify the information present in the *Subject, Student,* and *Teacher Related Factors* tables the procedures already described above were followed. Applying the algorithm presented in [9] to the table or relation's fields that make the knowledge base for QEGQL (Fig. 2), and looking to the DoC_s values obtained as described in [9], it is possible to set the arguments of the predicate *quality_evaluation_of_g(eneral)_c(hemistry)_l(earning)* (*quality$_{gcl}$*) referred to below, that also denotes the objective function with respect to the problem under analyze.

$$quality_{gcl}: Stud_{ent\ Issues}, Teacher_{Issues}, Subj_{ect\ Issues} \rightarrow \{0,1\}$$

where 0 (zero) and 1 (one) denote, respectively, the truth values *false* and *true*. Exemplifying the application of the algorithm presented in [9] to a term (clause) that presents feature vector ($Stud_{ents\ Issues} = 0.71$, $Teacher_{Issues} = 0.25...0.56$, $Subj_{ect\ Issues} = \perp$), one may have:

$$\{ \ \neg quality_{gcl} \left((QoI_{Stud}, DoC_{Stud}), (QoI_{Teacher}, DoC_{Teacher}), (QoI_{Subj}, DoC_{Subj}) \right)$$

$$\leftarrow not\ quality_{gcl} \left((QoI_{Stud}, DoC_{Stud}), (QoI_{Teacher}, DoC_{Teacher}), (QoI_{Subj}, DoC_{Subj}) \right)$$

$$quality_{gcl} \underbrace{\left((1, 1), \quad (1, 0.95), \quad (1, 0) \right)}_{attribute's\ quality-of-information\ and\ respective\ confidence\ values} :: 1 :: 0.65$$

$$\underbrace{[0.71, 0.71]\ [0.25, 0.56]\ [0, 1]}_{attribute's\ values\ ranges}$$

$$\underbrace{[0,1] \qquad [0,1] \qquad [0,1]}_{attribute's\ domains}$$

$\} :: 1$

4.1 Artificial Neural Networks

The model presented previously shows how the information comes together and how it is processed. In this section, a data mining approach to deal with the processed information is considered. A hybrid computing approach was set to model the universe of discourse, where the computational part is based on ANNs, which are used not only to structure data but also to capture the objective function's nature (i.e., the relationships between inputs and outputs) [10].

Now, looking at Fig. 3, we shall see a case being submitted to a QEGQL's assessment (its QoI's and DoC's values stand for the inputs to the ANN). The output is given in terms of a QEGQL's value and the degree of confidence that one has on such a happening. In this study 127 responses to the questionnaire were considered (i.e., one hundred and twenty seven terms or clauses of the extension of predicate quality$_{gcl}$ were considered). To ensure statistical significance of the attained results, 30 (thirty) experiments were applied in all tests. In each simulation, the available data was randomly divided into two mutually exclusive partitions, i.e., the training set with 67% of the available data, used during the modeling phase, and the test set with the remaining data (i.e., 33%), used after training in order to evaluate the model performance and to validate it. The back propagation algorithm was used in the learning process of the ANN. As the output function in the pre-processing layer it was used the identity one, while in the other layers it was used the sigmoid one.

Table 1 presents the coincidence matrix of the ANN model, where the values presented denote the average of 30 experiments. A glance at Table 1 shows that the model accuracy was 94.2% for the training set (82 correctly classified in 87) and 92.5% for test set (37 correctly classified in 40). Thus, the predictions made by the ANN model are satisfactory, attaining accuracies higher than 90%.

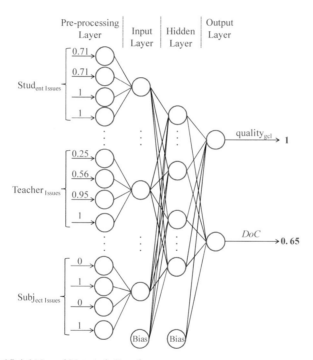

Fig. 3 The Artificial Neural Network Topology.

Table 1 The coincidence matrix for ANN model.

Target	**Predictive**			
	Training set		Test set	
	True (1)	False (0)	True (1)	False (0)
True (1)	62	2	29	1
False (0)	3	20	2	8

5   Conclusions

A QEGQL performance measurement is not only an inestimable practice, but something of utmost importance in a Higher Education context. The problems faced by contemporary society require from the Higher Educational Institutions the highest standards of quality in training future professionals. To meet this challenge it is necessary that the educational practice and the simultaneous evaluation of its impact on the students' learning process be intertwined. However, it is difficult to assess the QEGQL since it is necessary to consider different variables and/or conditions with complex relations entwined them, where the data may be incomplete, contradictory, and even unknown. This approach not only allows for the assessment of QEGQL but it also permits the estimation of a measure of confidence associated with such an evaluation. In fact, this is one of the added values of this method that arises from the complementarity between Logic Programming (for knowledge representation and reasoning) and the computing process based on ANNs (selected due to their dynamics like adaptability, robustness, and flexibility). Furthermore, this new methodology for problem solving may be used to assess the quality of learning in other subjects since the analyzed data do not refer explicitly to the contents taught, but how the learning process is organized.

Acknowledgments This work has been supported by COMPETE: POCI-01-0145-FEDER-007043 and FCT – Fundação para a Ciência e Tecnologia within the Project Scope: UID/CEC/00319/2013.

References

1. Coe, R., Searle, J., Barmby, P., Jones, K., Higgins, S.: Relative difficulty of examinations in different subjects, Report for SCORE – Science Community Supporting Education (2008). http://www.cem.org/attachments/score2008report.pdf
2. Rodríguez, R.M., Corona, L.B., Ibáñez, M.V.: Cooperative learning in the implementation of teaching chemistry (didactic instrumentation) in engineering in México. Procedia – Social and Behavioral Sciences **174**, 2920–2925 (2015)
3. Osma, I., Radid, M.: Analysis of the Students' Judgments on the Quality of Teaching Received: Case of Chemistry Students at the Faculty of Sciences Ben M'sik. Procedia – Social and Behavioral Sciences **197**, 2223–2228 (2015)
4. Ďurišová, M., Kucharčíková, A., Tokarčíková, E.: Assessment of higher education teaching outcomes (Quality of higher education). Procedia – Social and Behavioral Sciences **174**, 2497–2502 (2015)
5. Neves, J.: A logic interpreter to handle time and negation in logic databases. In: Muller, R., Pottmyer, J. (eds.) Proceedings of the 1984 Annual Conference of the ACM on the 5th Generation Challenge, pp. 50–54. Association for Computing Machinery, New York (1984)
6. Cortez, P., Rocha, M., Neves, J.: Evolving Time Series Forecasting ARMA Models. Journal of Heuristics **10**, 415–429 (2004)

7. Kakas, A., Kowalski, R., Toni, F.: The role of abduction in logic programming. In: Gabbay, D., Hogger, C., Robinson, I. (eds.) Handbook of Logic in Artificial Intelligence and Logic Programming, vol. 5, pp. 235–324. Oxford University Press, Oxford (1998)
8. Machado, J., Abelha, A., Novais, P., Neves, J., Neves, J.: Quality of service in healthcare units. In: Bertelle, C., Ayesh, A. (eds.) Proceedings of the ESM 2008, pp. 291–298. Eurosis – ETI Publication, Ghent (2008)
9. Fernandes, F., Vicente, H., Abelha, A., Machado, J., Novais, P., Neves, J.: Artificial neural networks in diabetes control. In: Proceedings of the 2015 Science and Information Conference (SAI 2015), pp. 362–370. IEEE Edition (2015)
10. Vicente, H., Couto, C., Machado, J., Abelha, A., Neves, J.: Prediction of Water Quality Parameters in a Reservoir using Artificial Neural Networks. International Journal of Design & Nature and Ecodynamics 7, 309–318 (2012)

A Study on Teaching and Learning the von Neumann Machine in a 3D Learning Environment

Maria Rosita Cecilia and Giovanni De Gasperis

Abstract The paper presents the design of three serious games for teaching the basis of the von Neumann's machine in a 3D environment. For this objective, the paper initially defines a framework useful to describe the design, then uses the framework to introduce the games. Furthermore, it presents a first prototype of one of the described games. It then describes the protocol that will used to evaluate the usability, proficiency and psychological effectiveness of such games, and ends with a brief discussion on the proposed study.

Keywords Virtual learning environment · Technology enhanced learning · Computer architecture

1 Introduction

Technology-enhanced learning is not a new concept. The use of technology to strengthen the student learning experience is a well established area of interest across all tiers of global education. Educators have always tried to integrate technology into the instruction process. However, innovations in content delivery, assessment methods, and adaptive learning have changed the way in which both teachers educate students and how students learn.

One of the innovations introduced in the TEL field has been the adoption of 3D technologies, as for enabling users to navigate, perform activities, and communicate among themselves at the same time, in a virtual space [1]. Another one has been

M.R. Cecilia
Department of Life, Health and Environmental Sciences, University of L'Aquila, L'Aquila, Italy
e-mail: mariarosita.cecilia@univaq.it

G. De Gasperis(✉)
Department of Information Engineering and Computer Science and Mathematics,
University of L'Aquila, L'Aquila, Italy
e-mail: giovanni.degasperis@univaq.it

© Springer International Publishing Switzerland 2016
M. Caporuscio et al. (eds.), *mis4TEL*,
Advances in Intelligent Systems and Computing 478,
DOI: 10.1007/978-3-319-40165-2_10

the use of games throughout the learning process. Games have recently attracted increasing interest among educators due to the growth of digital gaming in children, teenagers and adults, and also because games facilitate engagement and motivation [2, 3]. Game-Based Learning is an instructional method that incorporates educational content or learning principles into games to engage learners [4]. When games are placed in the context of learning (or training and marketing), they fall into the context of the called "serious games" [5]. Nevertheless, to become a valuable learning experience, serious games must merge a sound theoretical pedagogical underpinning, flanking the classical motivational and engaging aspects of games design.

In such a context, several pedagogical frameworks in a TEL context have been explored, e.g., Situated Learning [6], Problem-Based Learning [7], Experiential Learning [8], as well as many game design frameworks, e.g., EMAPPS [9] and TERENCE [10].

In the paper, we initially present the design of three serious games, developed for a 3D environment, for teaching and learning the basics of the von Neumann machine, for students of the first year of a medical degree and third year of Human Studies. For these games, their pedagogical basis is also discussed (section 2). The paper then continues by presenting the system, the first prototypes (section 3) and the planned experiment in which we aim at evaluating the usability of games and the preliminary foreseen psychological improvements (section 4). The paper then ends with a discussion about the future plans (section 5).

2   Design

2.1   Pedagogical Underpinnings

In recent years, a growing body of scientific studies has focused on the importance of underpinning serious game design and Game-Based Learning strategies with established instructional strategies and pedagogical theories [11]. Accordingly, a serious game design must be unquestionably underpinned by a sound pedagogical framework. In our study, the pedagogical approach is the well-known Problem-Based Learning [7], based on the following key aspects:

1. Knowledge is related to an experiential learning and it develops in response to learners' problem-solving actions [12, 13]. On the other hand, students need to be engaged in doing, rather than passively engaged in receiving knowledge [14].
2. Instruction is a process of supporting knowledge, rather than a process of communicating knowledge; teacher is a tutor instead of expert [12]. The focus shifts from teaching (via didactic instruction) to student learning via active and independent participation in problem-solving activities [15].
3. Learners are engaged in authentic and contextualized problems as near as possible to real life, in order to understand and solve them in a specific contest, and at the same time in order to stimulate and transfer problem-solving behaviors to real-life problems [15, 16].

Table 1 The game framework used in our study

Name	name of the game	
Goal	define what do you want the players to learn	
Instructions	instructions concerning the game, for players: specific to the game instance, motivational, concerning the rules	
Gameplay	3D Playing field	describe where the game takes place and what the player sees
	Interaction model	describe how the player interact with the world
	Challenges	describe the rules, obstacles and clues
Mechanics	Victory condition	describe how a player wins the game
	Loss condition	describe how a player loses the game
	Progress towards victory	write how the player can understand his/her progress towards the victory
Device	list of all devices available for the game	

Based on Problem-Based Learning Theory, our design strategy proposes different missions which the learner/player must strive to accomplish [17, 18]. Accordingly, the general game can be seen as a complex problem, comprising multiple goals [19, 20]. Each mission requires specific skills: (a) problem definition and formulation; (b) generation of alternative solutions; (c) decision-making, (d) solution implementation and verification [21]. Accordingly, the students will experiment their knowledge via an active and independent participation in problem-solving activities.

2.2 Framework

The subsection focuses on a framework in which we place the design of the three serious games used in our study. The framework starts from the EMAPPS [9] and TERENCE [10] frameworks, and is presented as a table (see Table 1).

Accordingly, in our framework, to design a game we must specify the game name, goal and instruction on how to play to get to the goal. Then, two big section have to be specified. The first is the gameplay, where we specify the playing field, how the player can interact with the objects located in the playing field, and which are the rules, obstacles and clues to win the game. The second section regards the game mechanics, that contains four parts: (i) the internal economy is where we specify who/how the objects are produced/used and the facets for the eventual adaptations, (ii) the victory condition is the exact definition of how a player wins the game, (iii) the loss condition, and (iv) how a player is able to understand his/her progresses towards the victory. A final section is also introduced, regarding the devices that can actually be used to interact with the games.

Table 2 The first game - identify the main components of the von Neumann machine

Name	Pick the computer parts	
Goal	Identification of ideal components of the von Neumann computer architecture	
Instructions	In front of you there are many objects that may be part or not of the well known von Neumann computer architecture. Select the ones that you think are the right ones.	
Gameplay	3D Playing field	The island, with buildings, among the many the computer science one. In the building, a first room. It contains a computer and electronics laboratory, with benches, chairs, electronics instrumentations and a set of objects of various electronic gears.
	Interaction model	Touching the objects and show if they are correct or not. Proximity to enable the internal scripts.
	Challenges	Recognize the right parts among a larger set that includes similar and look-like objects.
Mechanics	Victory condition	All correct objects are recognized.
	Loss condition	The number of wrong touched objects overwhelms the number of correct ones.
	Progress towards victory	Number of correct object touch events over the total number of right objects. Touching the wrong object reduces the score. Touching the correct object will increase it.
Device	PC, virtual world viewer	

Table 3 The second game: guided assembly: knowing the functions of each component of the von Neumann machine

Name	Build the machine	
Goal	Build the von Neumann architecture in the virtual world	
Instructions	In front of you there are many objects that may be part or not of the well known von Neumann computer architecture. Select the ones that you think are the right ones, and put them in their appropriate component slot.	
Gameplay	3D Playing field	The island, with buildings, among the many the computer science one. In the building, a first room. It contains a computer and electronics laboratory, with benches, chairs, electronics instrumentations and a set of movable objects.
	Interaction model	Touching each of the movable objects and touching a destination, the object will move the object in mid air to try to fit the indicated slot. If it does not fit, the object will go back where it started. Proximity to activate the internal scripts.
	Challenges	Recognize the right parts among a larger set that includes similar and look-like objects and how their fit into the final assembly.
Mechanics	Victory condition	All slots are filled in with the right components.
	Loss condition	Timeout without filling all the slots.
	Progress towards victory	Number of right object/slot combination over the total number.
Device	PC, virtual world viewer	

Table 4 The third game - interacting with the von Neumann machine: from the components to the general functioning

Name	Engage with the machine	
Goal	Understand how each component relates to the others, how information flows along the von Neumann architecture.	
Instructions	You see if front of you the assembled von Neumann computer architecture. You now have to interact with it by using the text chat, giving commands to each part so that the machine does an actual information processing, letting the information flow through all the components.	
Gameplay	3D Playing field	The island, with buildings, among the many the computer science one. In the building, a first room. It contains an almost empty computer room, with science fiction look and feel, with the 3D architecture of the machine in the middle.
	Interaction model	Text chat, touch, proximity position of the avatar.
	Challenges	Giving the right command to each component so that it is able to compute and/or input/output some data.
Mechanics	Victory condition	All components have been used at least once to compute a given expression.
	Loss condition	Not being able to give all the commands to the components to compute the given expression before a timeout.
	Progress towards victory	Number of correct commands, number of used components. Result of the computation. How many components are used to compute. The right sequence of activation of components in the CPU read/write cycle. Wrong commands reduce the score.
Device	PC, virtual world viewer	

2.3   Game Instances

In the following subsections, we describe the three games, according to the afore-mentioned framework.

1. Identify the main components of the von Neumann machine (see table 2)
2. Guided assembly - knowing the functions of each component of the von Neu-mann machine (see table 3)
3. Interacting with the von Neumann machine: from the components to the general functioning (see table 4)

3   Implementation

3.1   Architecture

The architecture used in our study mainly relies on server-side back-end software processes. The OpenSimulator main process is a MONO C# application that communicates with the client side virtual world viewer through a openmetaverse protocol over HTTP/UDP, as in [22]. It also uses the SQL server process as persistence memory of the world objects and properties, including internal objects scripts. Some of those scripts contains the games' logic and gateways to the redis event server that constitutes the adaptive system memory. The player is only using the virtual world viewer, among one of the existing open source package (FireStorm, Singularity or KoKua viewers) that encapsulates all the rendering, 3D data and interaction events.

3.2   Prototype

A prototype of game number 3 "Engage with the machine" has been implemented as reported by [22], upgraded with the redis server and its related LSL scripts as adaptive engine. In particular the LSL redis scripts log all the user activity triggered by the events of game interest (touch, chat, proximity). It also updates redis sets with abstract items representing achieved results so that the score can be easily calculated with cardinality of those sets. Few screenshots are available in Figure 1.

 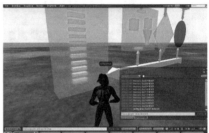

Fig. 1 Screenshots of the prototype

4   Evaluation

4.1   Experiment Design

The design of the experiment is a transversal study, for exploratory purpose, i.e., an observational study in which the exposure to a specific factor (i.e., the gaming activity) and condition (i.e. problem-solving skills) is determined at the same point in time in an exposed population wrt a control [23].

The aim of our study is threefold. First, to evaluate the pedagogical effectiveness of the system. Second to explore the psychological effectiveness of the system. Third, to assess the system usability. This leads to three corresponding research hypotheses. After the usage of the system the expected result shall be:

1. As for the pedagogical viewpoint, an improvement of the learners' proficiency;
2. As for the psychological viewpoint, a positive change in students' strategic knowledge and in their emotional attitude,
3. As for the usability viewpoint, an engagement of students giving feedback about the overall game playability

The preliminary results will allow us to understand how to improve our games.

The following subsection describes the details of the experiment in which the usability and psycho-pedagogical effectiveness will be evaluated.

4.2   Procedures

Potential participants will be students from the University of L'Aquila, attending the first year of Medicine and Surgery degree course and the third year of Philosophy and Communication Theory Processes degree course. The experiment is depicted in Figure 2.

In the first part of the study, all students will be divided into two groups respective to their courses, i.e. the Health Informatics course at the Medical School and Foundation of Computer Science course in Human Studies, concerning the basic knowledge of the von Neumann machine. It is worth noting that this part of experiment is already completed: all students took a 2h course unit on the basics of von Neumann machine during the month of November 2015. The second part of the experiment will be presented as a supplementary didactic activity, organized as follows. The games will be shown by the teacher that will explain how to interact with them. All participants will be randomly assigned to either Group 1 (G1) or Group 2 (G2). In a first step, only G1 will play with the games. The time for the students to complete all games will be at least 5 hours, which are needed to become acquainted with the 3D environment, move within the playing field, interact with the objects disseminated in the 3D world, and solve the puzzles that represent the final goals of the games. In a second step, both G1 and G2 will complete the proficiency and psychological tests, described in 4.2.1 and 4.2.2 respectively. It should take about 1 hour. In a third step, only G1 will perform the UX tests, while G2 will use the software. At the end of learning activity, also G2 will complete the UX tests described in Section 4.2.3.

Fig. 2 The experiment

The pedagogical, psychological and usability goals, as well as the outcome to assess are discussed separately. This distinction increases clarity, and therefore better comprehension about the related research findings.

Examine Students' Proficiency (Pedagogical Outcome). As described above, from the pedagogical viewpoint, the expected result will be the improvement of the learners' proficiency, i.e., the students' learning achievements about the principles of von Neumann machine after the usage of system. An Ad-Hoc Achievement Test will be used to measure students' achievement in learning. It will be developed based on the content of Informatics course by two experienced teachers in this field. In particular, this test will be used to assess the students' knowledge on von Neumann machine. We will evaluate proficiency comparing skill-based learning outcomes of G1 and G2 during the second step, when only G1 will have played with the games. More precisely, we will examine differences in proficiency between a group of students who will use the system (G1) vs. a group of learners who will not (G2), on a Ad-Hoc measure of academic competence about the principles of von Neumann machine. The expected outcome is that G1 will have higher scores than G2.

Examine Students' Strategic Knowledge (the First Psychological Outcome). As for the psychological viewpoint, the expected result will be a positive changes in students' strategic knowledge, i.e. the learners' study strategies and in their approach to study. Strategies may be defined as "goal-directed operations employed to facilitate task performance" [24]. Strategies are strongly related to problem-solving skills, e.g. they allow generating solutions to problems, they are potentially conscious and controllable, but they can be also automatic [25].The Study Strategies Questionnaire will analyze students' beliefs about a specific strategy (functional or dysfunctional for learning, e.g. mapping or diagrams to draw connections and show relationships between idea; make summary notes on the important concepts; integrate new information and knowledge; etc.) and its actual use. The Approach to Study Questionnaire will give information on students' working method and their approach to the study, in particular, their ability of organization, processing, self-evaluation, preparation for a test and metacognitive sensitivity. These questionnaires are part of Abilities and Motivation to Study Battery [26]. We will evaluate strategic knowledge comparing cognitive and metacognitive outcomes of G1 and G2 during the second step, when only G1 will have played with the games. More precisely, we will examine differences in strategic knowledge between a group of students who will use the system (G1) vs. a group of learners who will not (G2), on specific measure of strategic knowledge. The expected outcome is that G1 will have higher scores than G2. However, considering the specific characteristics of strategic knowledge, we do not expect significant differences between G1 and G2.

Examine Students' Emotional Attitude (the Second Psychological Outcome). We will evaluate emotional attitude comparing levels of self-efficacy, anxiety and resilience of G1 and G2 during the second step, when only G1 will have played with the games. Self-efficacy determines what activities people participate in, how much effort they will invest, how long they will persist to over-come challenging situations [27]. On the other hand, anxiety [28] and resilience[29] are critical to academic success. The Anxiety and Resilience Questionnaire will investigate emotional attitude toward their academic failure and success. This questionnaire is part of Abilities and Motivation to Study battery [26]. The General Self-Efficacy Scale, Italian version by Sibilia et al., [30], will evaluate students' belief in their ability to complete tasks and reach goals. Emotional attitude is not directly related to problem-solving skills, but it influences how the learners approach a problem. So the expected outcome will be that G1 will have higher scores than G2 about self-efficacy and resilience and lower about anxiety. However, considering the indirect relation with problem-solving strategies, we do not expect significant differences between G1 and G2.

Examine the Usability of the System. As for the usability, we will follow a quantitative approach [31], i.e., a set of UX metrics like the single ease question, time on task and System Usability Scale [32],which are easy to submit and fast to be collected. Student interactions events are recorded via a centralized log server that stores timestamps generated from the virtual world while students interact with virtual objects, such as proximity, click, chat and collision events.

In summary, the following research question will be explored:

– Examine differences in

1. Skill-based learning outcome of both G1 and G2
2. Cognitive and metacognitive learning outcomes of G1 vs G2
3. Affective learning outcomes of G1 vs G2

– Examine the usability of the system in all groups

5 Discussion

The paper presented a study in the threefold context of TEL, 3D and games. It presented the design, a first prototype and the proposed experiment to assess the effectiveness of our approach. The research group will focus in the coming years to introduce a general purpose 3D training session editor based on UML like MAS-CARET [33]. The motivations are twofold: (i) it already produces Unity learning objects suitable for mobile devices, (ii) in future it could also be integrating Open-Simulator since the base code is C#.

References

1. Maher, M.L., Skow, B., Cicognani, A.: Designing the virtual campus. Design Studies **20**(4), 319–342 (1999)
2. Connolly, T.M., Boyle, E.A., MacArthur, E., Hainey, T., Boyle, J.M.: A systematic literature review of empirical evidence on computer games and serious games. Computers & Education **59**(2), 661–686 (2012)

3. Boyle, E.A., Hainey, T., Connolly, T.M., Gray, G., Earp, J., Ott, M., Lim, T., Ninaus, M., Ribeiro, C., Pereira, J.: An update to the systematic literature review of empirical evidence of the impacts and outcomes of computer games and serious games. Computers & Education **94**, 178–192 (2016)
4. Tsai, C.H., Kuo, Y.H., Chu, K.C., Yen, J.C.: Development and Evaluation of Game-Based Learning System Using the Microsoft Kinect Sensor. International Journal of Distributed Sensor Networks (2015)
5. Johnson, W.L., Vilhjalmsson, H.H., Marsella, S.: Serious games for language learning: how much game, how much AI?. In: Proceedings of the 2005 Conference on Artificial Intelligence in Education: Supporting Learning Through Intelligent and Socially Informed Technology, vol. 125, pp. 306–313 (2005)
6. Krumsvik, R.J.: Situated learning and teachers? Digital competence. Education and Information Technologies **13**(4), 279–290 (2008)
7. Hmelo-Silver, C.E.: Problem-based learning: What and how do students learn? Educational Psychology Review **16**(3), 235–266 (2004)
8. Kolb, D.: Experiential Learning as the Science of Learning and Development. Prentice Hall, Englewood Cliffs (1984)
9. Davies, R., Krizova, R., Weiss, D.: eMapps.com: Games and Mobile Technology in Learning, pp. 103–110. Springer, Heidelberg (2006)
10. Cofini, V., De La Prieta, F., Di Mascio, T., Gennari, R., Vittorini, P.: Design Smart Games with requirements, generate them with a Click, and revise them with a GUIs. ADCAIJ: Advances in Distributed Computing and Artificial Intelligence Journal **1**(3), 55–68 (2013)
11. Kebritchi, M., Hirumi, A.: Examining the pedagogical foundations of modern educational computer games. Computers & Education **51**(4), 1729–1743 (2008)
12. Uden, L., Beaumont, C.: Technology and Problem-Based Learning. Information Science Reference, Hershey (2006)
13. Savin-Baden, M., Howell Major, C.: Foundations of Problem-Based Learning. Open University Press, Berkshire (2004)
14. Fosnot, C.T., Perry, R.S.: Constructivism: A psychological theory of learning. Constructivism: Theory, perspectives, and practice **2**, 8–33 (1996)
15. Savery, J.R., Duffy, T.M.: Problem based learning: an instructional model and its constructivist framework. In: Wilson, B. (ed.) Constructivist Learning Environments: Case Studies in Instructional Design. Educational Technology, Upper Saddle River (1996)
16. Delisle, R.: How to use Problem-Based Learning in the classroom. Association for Supervision and Curriculum Development, Alexandria (1997)
17. Kiili, K.: Digital Game-Based Learning: Towards an experiential gaming model. The Internet and Higher Education **8**(8), 13–24 (2005)
18. Kiili, K.: Foundation for Problem-Based Gaming. British Journal of Educational Technology **38**(3), 394–404 (2007)
19. Van Eck, R.: Digital Game-Based Learning: It's not just the digital natives who are restless. EDUCAUSE Review **41**(2), 16–30 (2006)
20. Tuzun, H.: Blending video games with learning: Issues and challenges with classroom implementations in the Turkish context. British Journal of Educational Technology **38**(3), 465–477 (2007)
21. Chang, E.C., D'Zurilla, T.J., Sanna, L.J.: Social Problem Solving: Theory, Research, and Training. American Psychological Association, Washington DC (2004)
22. De Gasperis, G., Florio, N.: Opensource gamification: a case study on humanities students learning computing architectures. In: Vittorini, P., Gennari, R. (eds.) Proceeding of ebuTEL 2013 - 3rd International Workshop on Evidence Based and User centred Technology Enhanced Learning. Springer, Berlin (2013)
23. Riffenburgh, H.: Statistics in Medicine, 3rd edn. Academic press (2012)
24. Harnishfeger, K.K., Bjorklund, D.F.: Children's strategies: a brief history. In: Bjorklund, D.F. (ed.) Children's Strategies: Contemporary Views of Cognitive Development, pp. 1–22. Erlbaum, Hillsdale (1990)

25. Pressley, M., Borkowski, J.G., Schneider, W.: Cognitive strategies: good strategy users co-ordinate metacognition and knowledge. In: Vasta, R., Whitehurst, G. (eds.) Annals of Child Development, vol. 5, pp. 89–129. JAL, New York (1987)
26. De Beni, R., Moé, A., Cornoldi C., Meneghetti, C., Fabris, M., Zamperlin, C., De Min Tona, G.: Abilitá e motivazione allo studio: Prove di valutazione e orientamento per la Scuola Secondaria di secondo grado e l'universitá. Nuova edizione [Abilities and Motivation to Study Battery: Evaluation and orientation testing for the second level of Secondary School and University. New Edition]. Erikson (2014)
27. Mun, Y.Y., Hwang, Y.: Predicting the use of web-based information systems: self-efficacy, enjoyment, learning goal orientation, and the technology acceptance model. International Journal of Human-Computer Studies **59**(4), 431–449 (2003)
28. Ma, X.: A meta-analysis of the relationship between anxiety toward mathematics and achievement in mathematics. Journal for Research in Mathematics Education, 520–540 (1999)
29. Martin, A.J., Marsh, H.W.: Academic resilience and its psychological and educational correlates: A construct validity approach. Psychology in the Schools **43**(3), 267–281 (2006)
30. Sibilia, L., Schwarzer, R., Jerusalem, M.: Italian adaptation of the general self-efficacy scale. Resource document. Ralf Schwarzer web site (1995) (accessed January 25, 2012)
31. Albert, W., Tullis, T.: Measuring the User Experience: Collecting, Analyzing, and Presenting Usability Metrics. Morgan Kaufmann Publishers Inc., San Francisco (2013)
32. Brooke, J.: SUS-A quick and dirty usability scale. Usability Evaluation in Industry **189**(194), 4–7 (1996)
33. Buche, C., Querrec, R., Loor, P.D., Chevaillier, P.: MASCARET: pedagogical multi-agents systems for virtual environment for training. In: Proceedings of 2003 International Conference on Cyberworlds, 2003, pp. 423–430. IEEE, December 2003

Identifying Essential Elements in Justifications of Student Drafts

José Rafael Hernández Tadeo, Aurelio López-López and Samuel González-López

Abstract The proposal draft is the first step to achieve a degree by students in many educational institutions. This proposal is transformed into a thesis after several revisions by an academic adviser. In addition, each proposal must comply with requirements of institutional guidelines. In this paper, we explore a learning approach to identify essential elements: importance, necessity, convenience, and benefits; that are expected to appear in a justification section of a proposal or thesis draft. We present a method based on a Language Model approach. Preliminary results show that the elements of necessity, importance, and benefits obtained acceptable results, considering that this task is complex for an academic adviser. The identification of convenience requires further improvement. A language model based on n-grams showed more consistent and better results than a model based on neural networks. Part of speech tagging contributes to improve results in both language model techniques.

Keywords Necessity · Importance · Convenience · Benefits · Justification · Student writing assessment

1 Introduction

Writing the first draft of a thesis is an arduous process for students and instructors. This document is transformed into the actual thesis document after several revisions by an academic adviser. The first proposal draft written by the student has usually

J.R.H. Tadeo · A. López-López
National Institute of Astrophysics, Optics and Electronics, Tonantzintla, Puebla, Mexico
e-mail: {rafaelhernandez,allopez}@inaoep.mx

S. González-López
Technological University of Nogales, Nogales, Sonora, Mexico
e-mail: sgonzalez@utnogales.edu.mx

© Springer International Publishing Switzerland 2016
M. Caporuscio et al. (eds.), *mis4TEL*,
Advances in Intelligent Systems and Computing 478,
DOI: 10.1007/978-3-319-40165-2_11

103

errors, derived from the inexperienced of the students in this process. Also, these documents must comply with the features established by institutional guidelines, such as format or structure. Our efforts aim to help students in preparing their drafts by examining the justification section in a proposal or thesis draft.

We explore the presence of four essential elements in the justification section [1] [2]: importance, necessity, convenience, and benefits. The justification section indicates the reasons to achieve the goal stated in the proposal or thesis. Below the elements are defined using questions that the student should answer when writing such section:

✓ Importance: What is transcendence of the research?
✓ Necessity: What are the essential aspects to be covered by the investigation?
✓ Convenience: What is the purpose of the research?
✓ Benefits: What are the enhancements obtained from the research?

In some cases the essential elements can appear at different levels in the paragraphs that make up a justification and can be visualized as four core dimensions as depicted in Figure 1:

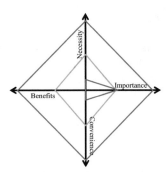

Fig. 1 Essential elements (dimensions) in a justification section.

These four selected dimensions can be expressed graphically with different colors (Figure 1). The blue color represents a justification section that complies satisfactorily with all dimensions, the color orange is a justification that shows less evidence of the four dimensions. Red color is a justification that presents only evidence of some dimensions, in this case mainly of importance and some convenience and necessity. Our method aims to find evidence of these dimensions in the Justification. Below, we provide a couple of examples of justifications that emphasizes the "importance" and "convenience" elements:

Currently, the University has approximately 3000 computers that need to be in perfect condition and must be used by the staff of this institution. It is important to use technology to help optimize resources, increase productivity, and improve the performance of support department.

The underlined sentence reveals the importance of the research, where the phrase "increase productivity" is related to the transcendence of the investigation. While the following paragraph serves to illustrate convenience:

Due to the increase in resolution of current images, object recognition requires higher processing, where the power of multiple processors or multiple cores can greatly reduce the execution time, that in the case of traditional sequential processing cannot be achieved. With the power of parallel processing, many processes can be fed to different processing units, resulting in a substantial improvement in runtimes, therefore, this is seen as a solution for the best performance of algorithms that require high computing power.

In the underlined text, the main purpose of the approach is provided. Similarly, the other essential elements can appear in a justification.

The task to be solved in this paper focuses on how to identify automatically the essential elements that are shown in Figure 1. An academic reviewer with experience in the academic texts evaluation can identify the paragraph or sentences that show the importance of the problem quickly, and even can determine the importance of justification in different domains. However, the scenario described above is computationally complex and represents a challenge [4]. This research seeks to establish the first steps to solve this problem, through natural language processing techniques, specifically training language models for the different elements.

This paper is organized as follow: section 2 presents related research, and in section 3 we include a description of data employed in our study, and then we delineate the methodology used in section 4. Furthermore, we show the experimental results in section 5. We conclude in section 6, detailing also work in progress.

2   Related Work

The evaluation of an academic paper is a complex task [4], and require that the academic reviewers provide a judgment based on their experience and domain knowledge. From a computational approach, it is intended that the judgment is consistent, so that the student can receive an homogeneous feedback. Similarly, the problem addressed in this work is complex to solve, as the elements of interest can be interpreted in different ways by reviewers or annotators.

In [5] Burstein proposed a machine learning approach to identify the thesis (main topic) and the conclusion section in student essays written on different topics. The results were favorable to identify the conclusion section, because the conclusion always was located at the end of the essay. In our work we assume that the justification section is previously bounded and the goal is identify internal expected elements (core dimensions) of such section.

The identification of features in academic writings is similarly found in the study of [6], in which the conclusions are analyzed using lexical features, cosine similarity and speculative terms. In our work, we conducted a characterization of the sentences of justification section to n-grams level and we used a language model to capture some patterns related to the elements.

Moreover, the study of Daunaravicius [7] analyzed papers to identify sentences that do not fit to the scientific writing genre. Some features addressed in the study were a formal notation and specific terminology used in the papers. Similarly, the solution proposed in our article focuses on the structure that each essential element of interest has, such as the importance or necessity.

3   Data Description

The corpus Coltypi[1] contains justifications of Graduate level: Master (MA) and Doctoral (PhD) degree; Undergraduate level that includes: Bachelor (BA) and Advanced College-level Technician (TSU) degree, where TSU corresponds to two year technical study program offered in some countries [3]. The corpus domain is computing and information technologies. Each item in Coltypi is a document that had been evaluated at some point by a reviewing committee. The total sentences extracted of justifications corpus were 1006, that includes 275 sentences related to some elements and 731 sentences without any element identified. Table 1 details the number of sentences describing particular elements.

Table 1 Experimental Collection

Essential element	Sentences	Train Set	Test Set
Importance	65	43	22
Necessity	55	36	19
Convenience	78	52	26
Benefits	77	51	26
Total	275		

Two annotators with academic experience marked sentences with the elements of interest, i.e. importance, necessity, convenience, and benefits. The paragraph sentences are part of each of the justifications of the collected corpus.

4   Methodology

The developed method seeks to capture the essential elements (core dimensions) within the paragraphs of the justification section. To perform the analysis,

---

[1] http://coltypi.org/

we employed a language model approach. The first step was to create a model for each element from the train set, then to each sentence of test set was calculated its perplexity value according to the corresponding language model. Two language models were trained: SRILM[2] (based on n-grams) y RNNLM[3] based on the implementation of recurrent neural networks by Tomas Mikolov [8].

The SRILM tool settings were taken as follows: 6-grams where used for all components and smoothing by the Ristad's natural method was used.

The RNNLM settings of the neural network were: No classes were used, the backtracking algorithm (6 times), and a hidden layer of 50 neurons was defined.

Text Preprocessing: Punctuation symbols and citations were removed, and numbers where replaced with the token "NUM". Then, the text was lemmatized with Freeling[4] with the default config file for Spanish, that also allowed to obtain the part-of-speech (POS) tags for the text.

Data Preparation: Each section was tested separately. For experiments, two subsets were obtained, one with 67% for training and the other with 33% for testing. The actual number of sentences in each set is given in Table 1. With the training set the model was built, for each of the essential elements (one with RNNLM and one with SRILM) with and without POS tags. Thereafter, each language model was applied to the test sets of all dimensions. Two hundred and fifty one sentences were randomly selected out of the 731 sentences without identified element, and added for testing purposes as negative samples.

The output of each language model tool is numerical, which represents perplexity that expresses the confidence of the sentence tested in the model. A low value of perplexity is higher confidence. The measure used to compare the similarity was "average perplexity per word" (ppl1), as detailed in Table 2:

Table 2 Languague model expression

$$ppl_1 = 10^{(\frac{-logprob}{words-OOVs})}$$
$$logprob = words - OOVs + sentences$$

words: identified words
OOVs: words out of vocabulary
sentences: total of sentences

To interpret the results of perplexity, the F-measure was calculated and the contingency matrix was created for each element. To build the matrix, we considered the annotated sentences by reviewer, that is, if a sentence obtained a low value of perplexity, this prayer was considered as a positive element (true positive). However, if the test sentence obtains a high value of perplexity means that sentence

---

[2] http://www.speech.sri.com/projects/srilm/
[3] http://rnnlm.org/
[4] http://nlp.lsi.upc.edu/freeling/

was not labeled in any of the categories (true negative) by the annotators. In addition, the threshold used to determine whether a sentence was positive or negative was estimated by averaging the perplexity value obtained by annotators, for each of the elements.

5   Experimental Results

First, we processed each of sentences of the test set in each of the four language model already built. We obtained the results showed in contingency matrix in Table 3 through 6, where the first and second tables includes the results produced by the n-gram language model and the third and fourth tables the results with the neural net-based language model:

Table 3 Contingency table for SRILM.

Dimensions	TP	FN	TN	FP
Importance	2	20	319	3
Necessity	1	18	307	18
Convenience	3	23	313	5
Benefits	1	25	287	31

Table 4 Contingency table for SRILM+POS tags.

Dimensions	TP	FN	TN	FP
Importance	2	20	322	0
Necessity	1	18	325	0
Convenience	2	24	316	2
Benefits	2	24	318	0

Table 5 Contingency table for RNNLM.

Dimensions	TP	FN	TN	FP
Importance	1	21	304	18
Necessity	1	18	301	24
Convenience	1	25	310	8
Benefits	1	25	273	45

Table 6 Contingency table for RNNLM+POS tags.

Dimensions	TP	FN	TN	FP
Importance	1	21	321	1
Necessity	1	18	322	4
Convenience	1	25	316	2
Benefits	1	25	302	16

Below in Table 5, we show the results in terms of weighted F-measure (*WFm*), which takes into account the number of elements per class, and is used to counteract the unbalanced classes, as expressed in:

$$WFm = FM_p * \left(\frac{P}{P+N}\right) + FM_n * \left(\frac{N}{P+N}\right)$$

where FM_p and FM_n are the F-measures for the positive and negative classes, and P and N are the number of positive and negative instances, respectively.

We can notice in Table 5 that the language model based on n-grams (SRILM) achieves better values than the second model, either with or without POS tags, identifying the elements of interest. Part of speech tagging provides additional information that allows to capture more relations in the language model, causing a noticeable   improvement in both kinds of language models.

Table 7 Essential element results (WFm)

Evaluated Element	Language Model Methods			
	SRILM	SRILM+POS	RNNLM	RNNLM+POS
Importance	0.913	**0.919**	0.883	0.910
Necessity	0.895	**0.925**	0.886	0.918
Convenience	**0.898**	**0.898**	0.882	0.892
Benefits	0.879	**0.902**	0.821	0.878

Regarding the elements of the justification, Necessity was more clearly identified by the language model than the rest, despite having less instances. However, both models also show a good discrimination capability for Importance. Benefits also reached acceptable value by SRILM but lower with RNNLM. Convenience, with higher number of instances, seems harder to identify, according to the results produced by both language model approaches.

6   Conclusions

The size of the experimental collection is relatively small, impacting on the trained language models. We have a plan to extend the corpus. We are also considering take into account word representations [9] that promise to improve the efficacy of the models.

It was expected that the language model based on recurrent neural networks have better performance, but the short length of each sentence may have caused that this did not happen. N-grams seem to work better on this task.

Besides POS information, we are exploring other features to improve the discriminative power of models.

Once we reach an acceptable classification level for the different elements, we will incorporate the language models in an online application to evaluate student drafts, providing feedback when needed, and helping in this way to both students and advisors.

Acknowledgements We thank the annotators that kindly reviewed the data collection. First author was supported by Conacyt México through scholarship 334435, while second author was partially supported by SNI, México.

References

1. Allen, G.: The graduate students' guide to theses and dissertations: a practical manual for writing and research. Jossey-Bass Inc. Pub., San Francisco (1976)
2. Hernández, R., Fernández, C., Batista, M.: Metodología de la investigación. Mc Graw Hill (2010)
3. González-López, S., López-López, A.: Colección de Tesis y Propuesta de Investigación en TICs: un recurso para su análisis y estudio. XIII Congreso Nacional de Investigación Educativa, p. 15 (2015)
4. Valenti, S., Neri, F., Cucchiarelli, A.: An Overview of Current Research on Automated Essay Grading. Journal of Information Technology Education **2**, 319–330 (2003)
5. Burstein, J., Marcu, D.: A Machine Learning Approach for Identification of Thesis and Conclusion Statements in Student Essays. Computers and the Humanities **37**, 455–467 (2003)
6. González, S., Bethard, S., López-López, A.: Identifying weak sentences in student drafts: a tutoring system. In: Methodologies and Intelligent Systems for Technology Enhanced Learning, vol. 292, pp. 77–85 (2014)
7. Daudaravicius, V.: Automated evaluation of scientific writing: AESW shared task proposal. In: Proceedings of the Tenth Workshop on Innovative Use of NLP for Building Educational Applications, pp. 56–63. Association for Computational Linguistics (2015)
8. Mikolov, T., Deoras, A., Kombrink, S., Burget, L., Cernocký, J.: Empirical evaluation and combination of advanced language modeling techniques. In: INTERSPEECH, pp. 605–608 (2011)
9. Turian, J., Ratinov, L., Bengio, Y.: Word representations: a simple and general method for semi-supervised learning. In: Proceedings of the 48th Annual Meeting of the Association for Computational Linguistics, pp. 384–394. Association for Computational Linguistics (2010)

Part V
Technology and Methodologies for Fostering Learning of Special Need Learners

Effectiveness and Usability of TERENCE Adaptive Learning System: The First Pilot Study with Children with Special Educational Needs

Maria Rosita Cecilia and Ferdinando di Orio

Abstract Early school leaving (ESL) is caused by a number of factors, including learning difficulties, social exclusion, lack of motivation, and scarce guidance and support from the school system. However, in Italy some progress was made to improve school quality and outcomes. In this context, we have run a project in an Italian primary and secondary school to test the psycho-pedagogical effectiveness and the usability of a compensatory technological tool, i.e., the TERENCE software, to support the learning process of students with Special Educational Needs (SENs), i.e., students that need particular attention due to biological, social and/or environmental factors. Specifically, we tested the TERENCE software to improve reading comprehension skills and stimulate learning motivation of 16 Italian students with SENs due to cultural, linguistic and socio-economic factors. Children used TERENCE in 8 interactive sessions, for 2.5 hours per session. Their comprehension skills were analyzed via MT standardized test at the beginning and at the end of the stimulation plan. Also usability data was gathered by adopting user-based methods like observational evaluation and semi-structured interviews. The results of the experiment were: (i) students' reading comprehension skills were significantly improved, (ii) students were highly involved and motivated and they learned how to interact with TERENCE very quickly. The pilot study with students with SENs suggested that TERENCE can be easily and effectively used to improve the learning experience of learners with SENs.

Keywords Effectiveness · Usability · Special educational needs

M.R. Cecilia(✉) · F. di Orio
Department of Life, Health and Environmental Sciences,
University of L'Aquila, L'Aquila, Italy
e-mail: {mariarosita.cecilia,ferdinando.diorio}@univaq.it

© Springer International Publishing Switzerland 2016
M. Caporuscio et al. (eds.), *mis4TEL*,
Advances in Intelligent Systems and Computing 478,
DOI: 10.1007/978-3-319-40165-2_12

113

1   Introduction

Early school leaving (ESL) is a complex phenomenon of disengagement which affect young individuals. ESL is caused by a number of factors, including learning difficulties, social exclusion, lack of motivation, and scarce guidance and support from the school system. ESL has been identified as a major obstacle to economic growth and unemployment in the European Union (EU), although the situation varies across countries [1]. This has triggered a strong commitment from all EU countries to reduce the rate of early leavers from education and training aged between 18-24 below 10%, by 2020. Progresses have been registered across the EU, although ESL still strongly persists in many EU countries, such as Italy. As of 2013, Italy registers a rate of early leavers of 17.0%, well above the EU average. Italy has committed to reduce the rate of early leavers to 16% by 2020, this target being considered more realistic for Italy, compared to the EU target of 10% [2].

However, in Italy some progress was made to improve school quality and outcomes, through new guidelines in this field. More precisely, the Italian Ministerial Directive "*Instruments of intervention for students with special educational needs and territorial organization for school inclusion*" (27 December 2012) has recently defined that a "*special*" attention is needed for a large number of students, because of biological, social and/or environmental differences [3]. The Directive recognizes that children with "*Special Educational Needs*" (SENs) are either:

- Children with handicap, Law 104/92 ("Framework Law for assistance, social integration and rights of the handicapped");
- Children with specific developmental disorders, e.g. specific learning disabilities, Law 107/10 ("New rules concerning specific learning disabilities in schools"), attention deficit hyperactivity disorder, language disorders, nonverbal skills deficit, motor skills disorder;
- Children with difficulties related to cultural, linguistic and socio-economic factors, such as the difficulty in learning the Italian language for non-native Italian speaking children.

With this Directive, teachers may analyze in a class council a specific learner's problem (divorce of parents, death of relatives, family problems, etc.) which prevents him/her to study and can decide to activate a personalized learning path. It may include the use of compensatory technological tool that can support the student for a period of time, until he/she overcomes the problems.

In this context we have run a project, together with an Italian primary and secondary school, the Comprehensive Institute "*G. Rodari*" in L'Aquila (Italy), that was funded under the Regional call "*Action 1. Friendly school – Activity A*", in P.O FSE 2007-2013, CRO Objective, Axis 4, Human capital - Special Project "*Friendly and inclusive school*" - "*Portfolio Project*" [4].

The project consisted in the use of a compensatory technological tool, i.e., the TERENCE software ("*An Adaptive Learning System for Reasoning about Stories with Poor Comprehenders and their Educators*") [5] for learners with SENs due to cultural, linguistic and socio-economic factors, such as the difficulty in learning the Italian language for non-native Italian speaking children (e.g., immigrant children). Children with SENs due to handicap or specific developmental disorders were not included in the project. The software is a web application designed by blending the "*User-Centered Design*" and "*Evidence-Based Design*" approaches, i.e., an adaptive learning system (ALS) designed to stimulate the reading comprehension skills of primary school hearing and deaf learners, in both Italian and English, being reading comprehension the most important key for school learning [6].

TERENCE proposes a customized stimulation, considering the specific characteristics, interests and abilities of each student. More precisely, the system presents to each learner a number of stories, collected in books and organized according to different levels of difficulty. Furthermore, a variety of games are provided, both to test the level of understanding of the stories (called smart games), to enjoy and to improve the computer skills needed to interact with the TERENCE system (called relaxing games).

Originally, TERENCE was developed for poor comprehenders, i.e., children that do not present any difficulty in decoding, but are not able to understand what they read [7]. Previous studies have demonstrated the TERENCE effectiveness and usability with poor comprehenders [8].

2   Objective

The aim of this project was to explore the effectiveness and usability of TERENCE with children with SENs due to cultural, linguistic and socio-economic factors. To this aim, we formulated the following research hypothesis:

i. The interaction with the TERENCE software could significantly improve the reading comprehension of students with SENs.
ii. The learning material and the Graphical User Interface (GUI) of the TERENCE system could be used by students with SENs to achieve specified goals (reading the stories and solving the related games) with effectiveness, efficiency and satisfaction in their specified context of use [9].

3   Materials and Methods

The experiment was designed as a pre/post, non-randomized study, involving learners with SENs, identified by their teachers. The project took place at school, as extra-curricular activity, during the school year 2014-2015 (from April to June).

The inclusion criteria were: primary and secondary school children with SENs due to cultural, linguistic and socio-economic factors, male and female, within 7-13 years old.

The exclusion criteria were: the lack of informed consent provided by their parents or legal guardians, complete lack of knowledge of Italian language, diseases/health conditions that did not allow the assessment of reading performances, handicap or specific developmental disorders.

Accordingly, the enrolled sample comprised 16 Italian students, aged between 9-13, of primary and secondary school of the Comprehensive Institute "*G. Rodari*" in L'Aquila (Italy) with learning difficulties due to cultural, linguistic and socio-economic factors, such as the difficulty in learning the Italian language for non-native Italian speaking children (e.g., immigrant children).

Of these 16 learners, 3 were excluded because their parents gave their consent to the stimulation plan, but not to the data collection. Therefore, data related to 13 learners was analysed. Eight learners were males and five females. All data was gathered anonymously.

The study was organized into three main phases:

- Phase 1: Pre evaluation phase - Learners were assessed via psychological tests to evaluate their initial comprehension level and to properly configure the TERENCE system, in particular its adaptive engine. All children began the stimulation phase with stories of the lowest level of difficulty. Furthermore, all children were gathered in a classroom and tested at the same time.
- Phase 2: Stimulation phase - Participants used the TERENCE software in 8 interactive sessions, for 2.5 hours per session, under the supervision of an expert psychologist and two teachers. In each session the children read a story, then played with games related to the content of the story (smart games) and eventually played with relaxing games. Each child used the TERENCE system via his/her device. Each session was run collectively in a dedicated classroom.
- Phase 3: Post evaluation phase - At the end, all children were re-tested like in Phase 1, so as to evaluate their final comprehension level.

Evaluating the psychopedagogical effectiveness of TERENCE consisted in comparing learners' reading comprehension skills at the beginning and at the end of the stimulation plan. Reading comprehension was assessed with Italian MT standardized tests: "*Prove di Lettura MT-2 per la Scuola Primaria*" ("*MT-2 Reading tests for Primary School*") [10] for 9-11 years old learners and "Nuove Prove di lettura MT per la Scuola Secondaria di I Grado" ("*New MT Reading tests for Secondary School, Grade I*") [11] for 11-13 years old learners. The child was asked to read a story and answer a set of multiple choice questions. A score was then calculated on the basis of the number of correct answers. A higher score corresponds to better comprehension. On the basis of their scores, children were classified as learners in normal condition or learners with reading comprehension difficulties. More precisely, children were grouped in clusters: "Need for immediate intervention" (NI), "*Attention is needed*" (AN), "*Sufficient performance*" (SP), "*Complete performance*" (CP). This was done by comparing children's scores with the conversion tables for the Italian population. Children with reading comprehension difficulties were assigned to one of the first two clusters (i.e. poor comprehenders).

Usability data were gathered adopting user-based methods like observational evaluation [12], semi-structured interviews [13]. The observational evaluation method was applied by observing the learners when interacting with the TERENCE system. Semi-structural interviews were carried out, with the aim of gathering the users' opinions on their motivation, satisfaction and interest, and with a particular focus on:

- The stories, whether appealing or not;
- The smart games, whether playful or too difficult;
- The relaxing games, whether sufficiently appealing or not.

In addition, the semi-structured interviews did not limit respondents to these set of pre-determined topics, but the learners were left free to discuss all aspects of their experience with the TERENCE system, in order to collect a broader range of usability information.

4   Results

Descriptive statistics were calculated for all MT clusters. A Wilcoxon signed-rank test was used to investigate the pre/post difference in comprehension skills [14].

The analysis of the MT test results showed that 4 students were poor comprehenders in the pre-evaluation phase and only one in the post evaluation phase, but 2 learners were not present during this second phase. However, they showed a sufficient performance in the pre evaluation one.

In summary, there was an increase of students with complete performance and a reduction of those with sufficient performance (see Table 1). More precisely, Wilcoxon signed-rank test indicated that the pre/post difference in MT clusters was statistically significant ($p < 0.03$).

Table 1 MT clusters in the pre evaluation phase (Pre) and in the post evaluation phase (Post)

	MT clusters							
	CP		SP		AI		NI	
	n	%	n	%	n	%	n	%
Pre	1	7.68	8	61.54	2	15.38	2	15.38
Post*	7	63.64	3	27.27	0	0.00	1	9.09

CP="Complete performance"; SP="Sufficient performance"; AI="Attention is needed"; NI="Need for immediate intervention".
*The numbers within the categories do not have the total of 13 due to missing data

The results showed that all learners gradually became proactive and interested in giving feedback concerning how to improve the TERENCE software. All usability findings are reported in details in the following specific categories

- *Reading activities.* Children reported that book titles were too small and some words were too difficult for younger learner (some learners with SENs did not speak Italian very well).
- *Smart games.* Learners generally liked the smart games, in particular causality games, which presented a fair level of difficulty. They also proposed to introduce a certain degree of competition in the games, for instance, by comparing the scores obtained in the smart games
- *Relaxing games.* Learners generally liked the relaxing games, however all students reported frustration when unable to complete the games due to the fixed time limit that is set in the TERENCE system to complete the games.

Learners were enthusiastic, focused, interested in what they were doing; they were highly involved and motivated, absorbed and immersed in all activities, and shown a strong will to succeed. Children liked to frequently interact with the TERENCE system and asked whether they could use it at home. Individuals that are highly motivated are more likely to engage in, devote effort to, and persist longer at a particular activity [16]. The teachers, who helped the psychologist in the project, highlighted the possibility of implementing the TERENCE system at school was an important opportunity. In particular, teachers were positively surprised of the impact of technology on student engagement in reading activities. Furthermore, the TERENCE system was evaluated as user-friendly by all learners, regardless of their level of experience regarding the use of technology. Before the pilot study, some children shown a lower affordance with gestures, mostly because many of them did not have any technological tools at home. However, all learners improved their computer skills. They learned to interact with the TERENCE system very quickly, without the need of constant support from the psychologist and/or the teachers.

5   Conclusion

The pilot study with students with SENs suggested that the TERENCE system can be easily and effectively used to improve the learning experience of students with SENs. It is worth highlighting that the current study suffers of two limits, i.e., the limited number of enrolled students and the lacking of a control group. Besides these limits, few preliminary results can be drawn.

Similarly as poor comprehenders [8], also learners with SENs may get benefit from the TERENCE system. Furthermore, they have a great need for this type of stimulation. Children improved their comprehension skills and they enjoyed testing what they had read, they were enthusiastic, focused, and engaged in all activities. This latter point was also an important finding, because students learn and retain the most from what Salomon calls "mindful" engagement [17]. Moreover, they cannot use a technological tool without being focused on learning activity [18]. Indeed, a tool may better facilitate the learning process if it is accepted by the learners [18]. In summary, motivation is a dimension that determines learning success, especially in complex e-learning environments [19]. We also noticed that the use of TERENCE system also created an exciting social and engaging classroom environment, which helped students to master traditional skills.

Moreover, teachers generally noted a high level of children engagement in reading activities. Teachers also suggested that the use of the TERENCE software could be included as both a curricular activity and extra-curricular activity in the upcoming school year 2015-2016 (from January to June), involving a larger number of students (35 learner). This could give as the possibility to address both the limitations before mentioned, i.e., including a larger sample and introducing a control group.

References

1. Bates, M., Saridaki, M., Kolovou, E., Mourlas, C., Brown, D., Burton, A., Battersby, S., Parsonage, S., Yarnall, T.: Designing location-based gaming applications with teenagers to address early school leaving. In: European Conference on Games Based Learning, p. 50. Academic Conferences International Limited (2015)
2. Colombo, M.: Early School Leaving in Italy. A Serious Issue, a Few 'Vicious Circles' and Some Prevention Strategies. Scuola Democratica 6(2), 411–424 (2015)
3. www.istruzione.it
4. www.regione.abruzzo.it
5. www.terenceproject.eu
6. McNamara, D.S.: Reading comprehension strategies: Theories, interventions, and technologies. Psychology Press (2012)
7. Nation, K.: Children's reading comprehension difficulties. In: Snowling, M.J., Hulme, C. (eds.) The Science of Reading: A Handbook. Blackwell Handbooks of Developmental Psychology, pp. 248–265. Blackwel Publishing, Oxford (2005)
8. Cecilia, M.R., Di Mascio, T., Tarantino, L., Vittorini, P.: Designing TEL products for poor comprehenders: Evidences from the evaluation of TERENCE. Interaction Design and Architecture(s) Journal - IxD&A 23, 50–67 (2014)
9. ISO, W.:9241-11. Ergonomic requirements for office work with visual display terminals (VDTs). The international organisation for standardization (1998)
10. Cornoldi, C., Colpo, G., Gruppo, M.T.: Prove di lettura MT-2 per la Scuola Primaria [MT-2 Reading tests for Primary School]. Giunti OS (2011)
11. Cornoldi, C., Colpo, G.: Nuove Prove di lettura MT per la Scuola Secondaria di I Grado. [New MT Reading tests for Secondary School, Grade I]. Giunti OS (2012)
12. Paramythis, A., Weibelzahl, S., Masthoff, J.: Layered evaluation of interactive adaptive systems: Framework and formative methods. User Modeling and User-Adapted Interaction 20(5), 383–453 (2010)
13. Russo, J.E., Johnson, E.J., Stephens, D.L.: The validity of verbal protocols. Memory and Cognition 17(6), 759–769 (1989)
14. Riffenburgh, H.: Statistics in Medicine, 3rd edn. Academic Press (2012)
15. Pope, C., Ziebland, S., Mays, N.: Analysing qualitative data. Qualitative Research in Health Care, 3rd edn., pp. 63–81 (2007)
16. Garris, R., Ahlers, R., Driskell, J.E.: Games, motivation, and learning: A research and practice model. Simulation and Gaming 33(4), 441–467 (2002)

17. Salomon, G.: On the nature of pedagogic computer tools: The case of the writing part-
    ner. Computers as Cognitive Tools, 179–196 (1993)
18. Jonassen, D.H.: Technology as cognitive tools: Learners as designers. IT Forum Paper
    1, 67–80 (1994)
19. Paas, F., Tuovinen, J.E., Van Merrienboer, J.J., Darabi, A.A.: A motivational perspec-
    tive on the relation between mental effort and performance: Optimizing learner in-
    volvement in instruction. Educational Technology Research and Development **53**(3),
    25–34 (2005)

The Use of Different Multiple Devices for an Ecological Assessment in Psychological Research: An Experience with a Daily Affect Assessment

Margherita Pasini, Margherita Brondino, Roberto Burro, Daniela Raccanello and Sara Gallo

Abstract Data gathering in psychological research is changing given the technological evolution and the availability of many different devices. However, very little research has been done to verify the validity of this kind of psychological measurements, even if a valid measure is the starting point for a valid research. Our study presents the results of a daily measurement of affect connected with the amount of time a person spends in a natural setting, considering the paradigm of restorative environments. High level of compliance was found from respondents, and also a good quality of the used measurements, in term of construct validity (using a Confirmatory Factor Analysis), invariance of the measure across days, and criterion related validity.

Keywords Innovative online and mobile assessment · Restorative environments · Positive affect · Multi-level confirmatory factor analysis

1 Introduction

Psychological research is often connected with the assessment of psychological constructs, and good quality of measurements is a fundamental requirement to increase internal validity, i.e. the extent to which a causal relation based on a study can be considered justifiable. The lack of systematic errors increase internal validity and

M. Pasini(✉) · M. Brondino · R. Burro · D. Raccanello · S. Gallo
Department of Philosophy, Education and Psychology,
University of Verona (Italy), Lungadige Porta Vittoria, 17, 37129 Verona, Italy
e-mail: {margherita.pasini,margherita.brondino,roberto.burro,daniela.raccanello}@univr.it,
sara.gallo@studenti.univr.it

© Springer International Publishing Switzerland 2016
M. Caporuscio et al. (eds.), *mis4TEL*,
Advances in Intelligent Systems and Computing 478,
DOI: 10.1007/978-3-319-40165-2_13

a usual source of systematic error in behavioural sciences could be the measurement instrument (scale) used to assess a psychological construct. A good scale has good psychometric properties, specifically validity and reliability. Nowadays, given the growth in technological developments, psychological constructs can be assessed using many different devices (personal computers, laptops, tablets, mobile phones). Even if many advantages can be connected with this possibility, there is a lack of scientific literature concerning the actual uses and the real effectiveness of these applications to increase data quality and consequently psychological research quality [16].

The use of these devices in the assessment could present various advantages. First, one advantage comes from the assessment mode, the self-administration in absence of a data collector. Data quality could be higher in self-administration modes, especially when the topic of interest is in some way sensitive [11]. In addition, people may tend to give more socially desirable responses in interviewers' administration than in self-administration [4].

Another advantage concerns the possibility, in contrast to what happens in laboratory studies, to record the construct of interest within the individual's environment, increasing ecological validity [14]. This possibility is extremely important in environmental psychology, in which research designs are used to explore human behaviour in relation to physical environment. The possibility to assess the constructs of interest when participants are in their real environment (allowing also to measure response latency) can increase the validity of the measures, with respect to the same assessment using lab studies and simulated environments, as pictures or virtual reality. Furthermore, assessing constructs as they naturally occur, permits to avoid some biases connected with retrospective self-report methods, for instance "peak" and "recency" effects [8].

Another advantage in using technological devices as personal computers, tablets or mobile phones to collect data in psychological research studies seems to be connected with higher and easier participation. Axinn, Gatny, and Wagner [1] found that allowing participants to switch modes in mixed-device survey research kept involved more participants compared to a web only approach. When questionnaires are dynamically programmed and suitable for completion on small devices, more people are inclined to use them for survey completion.

Even if this kind of assessment is becoming more and more frequent, the current state of knowledge about the dynamics of taking surveys on mobile devices is not as advanced as necessary, and more scientific literature is needed to evaluate some methodological aspects. For instance, it could be interesting to verify the participants' compliance. Another critical issue concerns the psychometrical properties of scales administered using these modes. A typical psychometric property of a scale consists in construct validity, i.e. the fact that the scale behaves as it should behave, respecting the underlying theory. Another important kind of validity which is usually verified is criterion-related validity, i.e the relation between the measure and another variable theoretically connected with the construct as a criterion variable.

Therefore, the present study aims to address these issues using a daily assessment of affect considering the paradigm of environmental psychology and restorative environments, in particular the "Stress Recovery Theory" [18] and the "Attention Restoration Theory" [9]. Taking into account these theoretical approaches, many research studies report that the exposure to natural environments produces positive changes in emotional states. In particular, experiencing natural environments is connected with health, well-being, and positive mood [3, 7, 17]. The huge amount of studies which report the positive effect of natural environment on affect lead us to choose this paradigm to verify the effectiveness of a multi-device assessment method. In the psychological literature, the two most general components characterizing affective experiences are Positive Affect, PA, and Negative Affect, NA (or Positive and Negative Activation, according to more recent formulations underlying their activation component) [20, 21]. Beyond being the structural dimensions of affect more frequently characterizing English mood terms, they represent the emotional dimensions underlying subjective well-being and they are strongly connected to personality factors [6, 19].

The aim of our study was to verify measurement quality of a daily affect assessment, related to how long participants had been in a natural setting or in a built setting during the day, along one week. This aim was explored in four steps:

- exploring which device was used, the perceived usability of each used device, and the participants' compliance along the week also related to the used device;
- testing the construct validity of the scale used to assess affect, verifying the theoretically based one-factor structure;
- testing the measurement invariance of the same instruments along the seven days in the week;
- testing the criterion-related validity using how long a participant had been in a natural environment as the criterion variable.

2   Method

2.1   Participants

The sample included 108 Italian first- and second-year undergraduate students enrolled at a Psychology degree (mean age: 25 year, SD=6.1; 95% female). Informed consent forms described the potential participants of the goals of the study and that they could stop their participation at any time during the study. To be included in the study, participants had to give their e-mail address. Data were collected between November 13 (Friday) and November 19 (Thursday), 2016. After data collection, all participants were informed in a public meeting about the results.

2.2   Material and Procedure

During the study week, the participants received an e-mail message every morning at 10, in which they were reminded to answer to an online questionnaire between 6 p.m. and midnight.

The online questionnaire was developed and administered using the Apsym-Survey Software (ApSS). ApSS is a LAMP[1] customization of the LimeSurvey open-source project that allows to create powerful online question and answer surveys that can work for a few or many simultaneous participants, close or far away, by means of a common internet browser. ApSS follows a Branching/Skip logic procedure, in order to result self-guiding for the participants who are responding. The main operating characteristics that ApSS offers are: 30 different question types, multi-lingual surveys, WYSIWYG (acronym for 'What You See Is What You Get') online-editor, possibility to integrate pictures and movies into a survey, anonymous/not-anonymous surveys, option to buffer answers to continue a survey at a later time, cookie or session based surveys, survey expiry dates, import/export functions for many types of file-extensions, descriptive statistical analyses. In particular, it allows the use of multiple different types of devices (computers, tablets, smartphones) for an ecological assessment due to its responsive-layout, i.e. a web design-approach oriented to provide a final adjusted viewing and interaction experience, comfortable reading and navigation with a minimum of resizing and scrolling across a wide range of devices.

In order to have a better user-experience, an Apsym-Survey Software application for Smartphone/Tablet (ApSS-app, for Android and IPhone/IPad devices) was developed using Apache Flex, an open-source application framework/cross-platform runtime system that allows to develop mobile applications by Action-Script (i.e., an object-oriented programming language).

The questionnaire focused on several measures, among which Positive and Negative Affect and time spent in physical environments. The kind of device used by participants and usability of each device were also monitored with some specific questions.

2.2.1   Positive and Negative Affect Schedule

Affect was measured using the italian version of the Positive and Negative Affect Schedule, PANAS [20], validated into Italian by Terracciano, McCrae, and Costa [15]. It's a 20-item scale presenting adjectives referred to two higher-order factors, Positive Affect (e.g., *active, enthusiastic, excited*) and Negative Affect (e.g., *afraid, upset, distressed*), with ten adjectives for each factor. Participants were asked to evaluate how much they had felt the state described by each adjective on

---

[1] LAMP is an model of web service solution stacks, acronym of the names of four original open-source parts: Linux, the well-known Unix-like computer operating system (OS), Apa-che HTTP Web-Server, MySQL relational database management system (RDBMS), and PHP server-side scripting language for web development.

a 7-point Likert-type scale (1 = *not at all* and 7 = *very much*), referring to the current day. Since natural environment should increase positive mood, only the Positive Affect dimension was considered in the present study.

2.2.2 Physical Environment

One item was used to assess the amount of time a person was in a natural environment during each day. The question concerns how long participants had been in a natural setting, and it was evaluated on a 5-point Likert scale (1 = never; 2 = until 30 minutes; 3 = from 30 minutes to 1 hour; 4 = from 1 to 2 hours; 5 = more than 2 hours).

2.2.3 Devices' Use and Usability

Which device has been used and the evaluation of usability of each device were assessed in the last-day survey, with some specific items.

2.3 Data Analyses

First, descriptive statistics were computed, to explore the use of the different devices and the compliance. We define compliance as answering to the questionnaire all the seven days. To test construct validity, Confirmatory Factor Analysis (CFA) was performed, to verify the mono-factorial structure of the scale. Due to the structure of the data, collected for seven days on the same sample, multilevel CFA was run using complex option of Mplus, version 6.11 [12]. We took into account the Chi-square ($\chi2$), the comparative fit index (CFI), the Tucker-Lewis index (TLI), the root-mean-square error of approximation (RMSEA), and the standardized root-mean-square residuals (SRMR), with CFI $\geq$ .90, RMSEA $\leq$ .08, and SRMR $\leq$ .11 as threshold values [2]. Measurement invariance was verified with a multi-group CFA. We checked for the model invariance across the seven days assessing the factor structure simultaneously for each day. Measurement Invariance (MI) analyses examined hypotheses on the similarity of the covariance structure across days by considering: (1) configural invariance, allowing all the parameters to be freely estimated; (2) metric invariance, requiring invariant factor loadings; (3) scalar invariance, requiring also invariant intercepts; and (4) uniqueness invariance, requiring invariant item uniqueness. Due to the small size of our sample, support for non-invariance required Δ CFI $\leq$ -.005, supplemented by Δ RMSEA $\geq$ .010 or SRMR $\geq$ .025, for testing metric invariance, and .010 or .005, respectively, for testing scalar and uniqueness invariance.

To test criterion-related validity, bivariate Pearson's correlation coefficient between positive affect and the amount of time spent in a physical environment during the seven days was used.

3 Results and Discussion

3.1 Use of the Devices and Compliance

Smartphone has been the most used device, with 77.8% of the participants using it at least one time, followed by notebook (41.7%), tablet (20.4%), PC desktop (15.7%) and mobile phone without touch screen (11.1%). When asked to estimate the percentage of use of a device during the week, mean percentage of use of smartphone was 78.8%, 34.4% for notebook, 17.2% for tablet, 14.4% for mobile phone, and 11.8% for PC desktop. Concerning usability, estimated on a 7-point Likert scale, all the devices were evaluated quite easy to use: notebooks and tablets were evaluated as the easiest devices, with 83.7% and 81.3% of the sample choosing the first two point of the scale ("not demanding at all", "low demanding"). A bit more demanding was considered the smartphone, with 52.9% of participants considering it "not demanding at all" or "low demanding".

Participants' compliance was quite high: 77.8% of participants answered each of the seven days. Considering the participants who did not comply, 70.8% did not answer once, 20.8% did not answer twice, and 8.3% of the sample did not answer for three days. Along days, compliance varied from 97.2% to 93.5%, and no week-end effect could be identified [5]. Compliance was also related to the used device: only 19% of missing was present in participants who declared to use the smartphone more frequently than the other devices.

3.2 Psychometric Properties of the PANAS Positive Affect Subscale

Since some adjectives of the scale were almost synonymous (enthusiastic and excited; concentrated and attentive), we decided to add two covariances between errors in the model. The model showed good fit indexes: $\chi2(33) = 151.56$, $p < .001$, CFI =.96, TLI = .94, RMSEA = .070, and SRMR = .036. All factor loadings were statistically different from 0, ranging from .48 to .80. This result confirms the theoretical one-factor structure of the Positive subscale of the PANAS administered using this mode.

3.3 Invariance of Measurement across Days

The results of the sequence of gradually more restrictive tests of MI supported the higher level of invariance, in particular: metric invariance (Δ CFI = .000; Δ RMSEA = .004), scalar invariance (Δ CFI = .002; Δ RMSEA = .002), and uniqueness invariance (Δ CFI = .005; Δ RMSEA = .008), across days. This means that the same factor structure of the scale, that is the theoretically based one-factor structure, was confirmed across the seven days.

3.4 Criterion-Related Validity

Bivariate Pearson's correlations were computed, day by day, between the extent of the time spent in a natural environment and the Positive Affect. Table 1 shows these correlations.

Table 1 Pearson's correlation coefficients (r) between time spent in a natural setting and Positive Affect, for the seven days

	Day 1	Day 2	Day 3	Day 4	Day 5	Day 6	Day 7
Correlations between time spent in a natural setting and Positive Affect	.21*	.21*	.26*	.21*	-.11	.20*	-.01

*$p < .05$.

Except for day 5 and day 7, for which the two variables were not related, in all the other five days we found significantly positive moderate correlations, confirming that a more intense positive affect corresponds to a higher amount of time one person spends in a natural setting.

4 Conclusions

Our research study aimed at investigating the measurement quality of a daily affect self-assessment using different multiple devices, in the participants' everyday life setting and not in a lab. The advantage to use this assessment mode is connected to the possibility to assess a psychological construct in an ecological situation. This is extremely important, for instance, in environmental psychology, that study the physical environment's impact on human perceptions and behaviours. In the present study the daily affect assessment was also related to the amount of time participants have been daily in a natural setting, considering the paradigm of environmental psychology and restorative environments as the theoretical frameworks [9, 18].

First, an interesting result concerns the most used device by the sample of young-adult participants: smartphone was the favourite device, and the other devices (in order: notebook, tablet, PC desktop, mobile Phone with no touch-screen) was clearly less used. This could depend on the age and educational level of participants: younger and educated population tend to use smartphone more than other populations, and more than other devices [10]. Despite the fact that this was the favourite device, smartphone's usability for this survey was considered lower than other devices' usability, such as notebook or tablet. This result suggests that more information should be gathered about the reasons of this lower perceived usability in the present case. Data also documented a high participants' compliance along the week (highest for smartphone users), suggesting that proposing an easy way for participants to respond to a survey in their everyday life setting could lead to a higher compliance and to lower sample mortality. To verify these speculations, future works should compare directly compliance investigated with the described modalities to compliance typical of more traditional data gathering instruments. Second, our analyses indicated the goodness of the structure of the PANAS Positive Affect subscale, supporting the construct validity of the measurement instrument to assess affect also when presented through a multi-device approach system. Third, measurement invariance of measures of positive affect along seven days in the week further supported their validity within multi-device surveys. Finally, the relation between natural environment and positive affect documents the criterion-related validity of the measurement of affect also in this mode; in addition, these findings contribute to extend current theoretical knowledge on the links between natural environments and affect for a specific sample.

A severe limitation of this study concerns the sociodemographic characteristics of the sample, which was mainly composed of female and young participants. It should be important to repeat the same research using samples more balanced for sex, and also involve older participants, to exclude a cohort effect. Future directions could also explore the measurement invariance across modes [22], to separately verify whether differences in the quality of measurements could be found considering the different devices. Also considering computer skills of participants could be interesting in future researches.

Relation between positive affect and amount of time spent in a natural setting was found, but in this study no attention has been paid to the perceived restorativeness of the natural setting, which should be the direct cause of positive affect: Assessing also this perception, using the same mode and a short instrument [13], could contribute to explain better the link between natural environment and positive affect.

Acknowledging limitations such as, for instance, the impossibility to separate the effect of different devices, or the non-balanced by sex and age sample, our work could be considered a first step to document how the growth in technological developments can be advantageous in terms of better data quality and related psychological research quality, considering the lack of scientific findings documenting it.

References

1. Axinn, W., Gatny, H., Wagner, J.: Maximizing data quality using mode switching in mixed-device survey design: Nonresponse bias and models of demographic behaviour. Methods, Data, Analyses **9**(2), 163–184 (2015)
2. Beauducel, A., Wittmann, W.W.: Simulation study on fit indexes in CFA based on data with slightly distorted simple structure. Structural Equation Modeling **12**, 41–75 (2005). doi:10.1207/s15328007sem1201_3
3. Berto, R.: The role of nature in coping with psycho-physiological stress: A literature review on restorativeness. Behavioural Sciences **4**(4), 394–409 (2014). doi:10.3390/bs4040394
4. Bowling, A.: Mode of questionnaire administration can have serious effects on data quality. Journal of Public Health **27**(3), 281–291 (2005)
5. Courvoisier, D.S., Eid, M., Lischetzke, T., Schreiber, W.H.: Psychometric properties of a computerized mobile phone method for assessing mood in daily life. Emotion **10**(1), 115–124 (2010)

6. Diener, E., Emmons, R.A., Larsen, R.J., Griffin, S.: The Satisfaction With Life Scale. Journal of Personality Assessment **49**, 71–75 (1985)
7. Hartig, T., Mitchell, R., de Vries, S., Frumkin, H.: Nature and health. Annual Review of Public Health **35**, 207–228 (2014)
8. Kahneman, D.: Objective happiness. In: Kahneman, D., Diener, E., Schwarz, N. (eds.) Well-being: The Foundations of Hedonic Psychology, pp. 3–25. Russell Sage Foundation, New York (1999)
9. Kaplan, S.: The restorative benefits of nature: Toward and integrative framework. Journal of Environmental Psychology **15**, 169–182 (1995)
10. Kim, Y., Briley, D.A., Ocepek, M.G.: Differential innovation of smartphone and application use by sociodemographics and personality. Computers in Human Behavior **44**, 141–147 (2015)
11. Kreuter, F., Presser, S., Tourangeau, R.: Social desirability bias in CATI, IVR, and web surveys: The effects of mode and question sensitivity. Public Opinion Quarterly **72**, 847–865 (2008)
12. Muthén, L.K., Muthén, B.O.: Mplus user's guide, 5th edn. Muthén & Muthén, Los Angeles (1998–2007)
13. Pasini, M., Berto, R., Brondino, M., Hall, R., Ortner, C.: How to measure the restorative quality of environments: The PRS-11. Procedia - Social and Behavioral Sciences **159**, 293–297 (2014)
14. Shiffman, S., Stone, A.A., Hufford, M.R.: Ecological momentary assessment. Annual Review in Clinical Psychology **4**, 1–32 (2008)
15. Terracciano, A., McCrae, R.R., Costa, P.T.: Factorial and construct validity of the Italian Positive and Negative Affect Schedule (PANAS). European Journal of Psychological Assessment **19**(2), 131–141 (2003). doi:10.1027//1015-5759.19.2.131
16. Toepoel, V., Lugtig, P.: Online surveys are mixed-device surveys. Issues associated with the use of different (mobile) devices in web surveys. Methods, Data, Analyses **9**(2), 155–162 (2015). doi:10.12758/mda.2015.009
17. Tyrväinen, L., Ojala, A., Korpela, K., Lanki, T., Tsunetsugu, Y., Kagawa, T.: The influence of urban green environments on stress relief measures: A field experiment. Journal of Environmental Psychology **38**, 1–9 (2014)
18. Ulrich, R.S.: Aesthetic and affective response to natural environment. In: Altman, I., Wohlwill, J.F. (eds.) Behavior and the Natural Environment, pp. 85–125. Plenum, New York (1983)
19. Watson, D., Clark, L.A.: On traits and temperament - General and specific factors of emotional experience and their relation to the 5-factor model. Journal of Personality **60**, 441–476 (1992)
20. Watson, D., Clark, L.A., Tellegen, A.: Development and validation of brief measures of positive and negative affect: The PANAS scales. Journal of Personality and Social Psychology **54**, 1063–1070 (1988)
21. Watson, D., Wiese, D., Vaidya, J., Tellegen, A.: The two general activation systems of affect: Structural findings, evolutionary considerations, and psychobiological evidence. Journal of Personality and Social Psychology **76**, 820–838 (1999)
22. Wong, C.K.H., Lam, C.L.K., Mulhern, B., Law, W.-L., Poon, J.T.C., Kwong, D.L.W., Tsang, J.: Measurement invariance of the Functional Assessment of Cancer Therapy— Colorectal quality-of-life instrument among modes of administration. Quality of Life Research **22**, 1415–1426 (2013). doi:10.1007/s11136-012-0272-x

Co-Robot Therapy to Foster Social Skills in Special Need Learners: Three Pilot Studies

Lundy Lewis, Nancy Charron, Christina Clamp and Michael Craig

Abstract Three studies are described in which co-robot therapy with the humanoid robot NAO is used to foster social skills in three subjects with special needs – two subjects with autism spectrum disorder and one subject with developmental delay, speech-language impairments, and tantruming/yelling. The social skills include imitation, attention, pro-social behavior, joint attention, turn-taking, and initiative. We discuss the base-line performance of the subjects in each study. The contributions of the paper are (i) the use of soft systems methodology to guide co-robot therapy over a multiplicity of incremental therapy sessions, (ii) studies in messy real-world environments, and (iii) the transition from a tele-operated robot to an intelligent agent-based robot.

Keywords Humanoid robot · Socially assistive robot · Methodology · Special needs learners · Intelligent agent

1 Introduction

A problem of increasing global concern is the dramatic increase of children with Autism Spectrum Disorder (ASD). Statistics in the US for 2010 identify 1 in 68 American children as on the autism spectrum, a ten-fold increase over 40 years [1]. A related problem is children with Specific Language Impairment (SLI), SLI is one of the most common childhood learning disabilities, affecting approximately 7 to 8 percent of children in kindergarten. Children with SLI have significant communication problems, which are also characteristic of most children with ASD [2].

L. Lewis(✉) · N. Charron · C. Clamp
Southern New Hampshire University, Manchester, NH, USA
e-mail: {l.lewis,n.charron,c.clamp}@snhu.edu

M. Craig
Sunset Heights Elementary School, Nashua, NH, USA
e-mail: craigm@nashua.edu

The original version of this chapter was revised: the correct figure has now been updated. The correction to this chapter is available at https://doi.org/10.1007/978-3-319-40165-2_21

© Springer International Publishing Switzerland 2016, corrected publication 2023
M. Caporuscio et al. (eds.), *mis4TEL*,
Advances in Intelligent Systems and Computing 478,
DOI: 10.1007/978-3-319-40165-2_14

Prior studies suggest that robots who interact socially with ASD children may prove to be useful tools for communication therapies, e.g. [3] compared subjects' interactions with a robotic dinosaur, a human, and touchscreen computer. Reference [4] examines the physical appearance of robots, the content and goals of human-robot interactions, types of evaluation studies, and data collection and analysis. Recent systematic reviews of the literature argue that co-robot therapy has potential to elicit different types of behaviors in ASD children of different ages and levels of ability but there is a need for studies that examine the incremental benefit of such interactions over well-defined participant groups, where directions for future research include (i) the extent to which the use of skills increase during subsequent interactions with humans, (ii) therapeutic protocols, i.e. the number, arrangement, and roles of participants during therapy, (iii) therapist-controlled robots without a robotic operator, and (iv) cognitive mechanisms that are affected by robot vs. human interactions [5,6,7]. The study in [8] focuses on the difficulty of collaboration between the robotics and clinical autism research communities.

In practice, co-robot therapy is a hard interdisciplinary problem requiring collaboration among parents, clinicians, speech and language pathologists (SLPs), special education (SPED) teachers, robot behavior designers, robot operators, and others [9]. We use Soft Systems Methodology (SSM) as an aid to facilitate collaboration [10]. Our experience with SSM suggests that it is useful for problems in human affairs for which there are multiple stakeholders with different methods, mindsets, and views of the problem and during which a notional system is discovered and improved upon iteratively [10,11]. In the studies described below, we use the NAO humanoid robot as an interactive partner with three subjects, also used successfully in a recent study showing that robot therapy can improve joint attention skills [12]. NAO is a two-foot tall ten pound humanoid robot with facilities for speech recognition, object recognition, mobility, and gesturing [13]. A useful feature of the NAO robot is its programmability as requirements change.

2   The First Study: Attention and Imitation

Subject 1 in the first study is a 6-year old male with mild ASD. According to his mother, his deficits in social communication cause noticeable impairments. He has difficulty initiating social interactions and shows atypical or unsuccessful response to social overtures of others.

The subject's mother collaborated with a special education (SPED) teacher in which it was decided to use the robot to teach the subject how to order a doughnut from a menu in a real doughnut shop. The SPED teacher then collaborated with the robot programmer to produce behaviors for a 1st therapy session. Three routines were implemented: (i) an introductory routine in which the robot engaged in introductions, calling the subject by name, (ii) an imitation routine where the robot asked the subject to imitate him, e.g. raising hands, waving, sitting, wiping forehead, and (iii) a routine in which the robot was programmed to take a menu, walk up to an imaginary counter, point to the menu, and ask for a doughnut and then ask the subject to do the same.

Two sessions took place in February 2014 at the subject's home. The robot was placed on the floor facing the subject, his siblings, and his parents. The routines were executed in the robot remotely by a robot operator. Fig. 1 provides context for the reader (the authors have obtained necessary clearance from parents). After the first session, the subject was taken to a doughnut shop to transfer the ordering skill to a real setting. He performed the task successfully, albeit softly and without confidence.

Fig. 1 Subject 1 on the left during the first therapy session

There are clear limitations to this first study. The result is motivating but it is anecdotal, i.e. a one-shot experiment lacking a formal research method and quantitative data. However, we learned robot therapy is an iterative process that consisting of three activities: session preparation, session execution, and post-analysis where post-analysis provides input for the next session. Further, we learned that it is useful to view robot therapy as comprising a system consisting of a number of components, e.g. human components, environmental components, and a robotic component, engaging in multiple activities. We used SSM to analyze the sessions retrospectively and to structure the evolution of subsequent studies. SSM is designed for difficult problems in human affairs in which a vague, notional system is discovered and then improved upon iteratively in a cycle of inquiry and discovery. Fig. 2 shows the methodology.

The methodology is cyclic. In the initiating cycle, our real-world problem situation was stated simply as "how to teach the subject how to order a doughnut in a real setting." The problem led to various purposeful activities in preparation for the therapy session, including collaboration between the SPED teacher and mother, collaboration between the teacher and the robot programmer, and the implementation/testing of behaviors in the robot, the actual 1st therapy session, and visiting the real doughnut store as a test. These activities are viewed as interactive subsystems which may produce emergent properties of the system as a whole. The 2nd cycle included these same activities plus the post-analysis of the session, resulting in the accommodation of modified behaviors for the 2nd session.

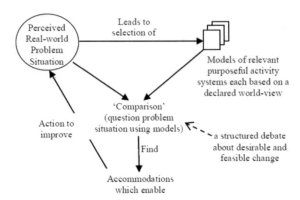

Fig. 2 The inquiring/learning cycle of SSM [11]

3 The Second Study: Joint Attention

A second study with two participants – Subject 2 and Subject 3 – took place in an elementary school during the Fall of 2014. Two subjects received therapy simultaneously in six 30-minute sessions in the office of their Speech and Language Pathologist (SLP). Due to absences, Subject 2 participated in five of the sessions and Subject 3 participated in three of the sessions. The SLP selected the subjects a priori. For each session, the robot was placed on a table in the SLP's office. The subjects and the SLP sat in chairs facing the robot. The robot was tele-operated by the robot operator to the side of the robot. Fig. 3 provides context.

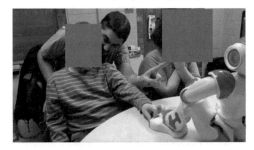

Fig. 3 Subject 2, SLP, Subject 3, and robot in the second study

Subject 2 is a 7-year old male in the 2nd grade. His SLP describes him thus: He shows deficits in maintaining attention, joint attention, and taking initiative. He participates in special and general education classes, receives direct SPED instruction and gets support from paraprofessionals. He participates in individual and group speech therapy sessions. He plays games, watches TV shows, and scripts them both physically and mentally, showing a strong cognitive inner life. He reads at 1st to 2nd grade level but comprehension is at kindergarten level.

Subject 3 is a 9 year-old female in the 3rd grade. Her SLP describes her thus: Her educational identification is Developmental Delay and Speech-Language Impairments. She participates in a special education classroom where she receives direct instruction from a special educator and support from a paraprofessional. She also receives individual and group speech therapy sessions. She enjoys playing with dolls and games as part of her Individualized Educational Program and she enjoys interacting with familiar adults. She has difficulties participating in activities appropriately due to interfering behaviors such as tantruming and yelling.

The focus of the therapy was to develop joint attention skills. In the beginning sessions, the robot introduced himself and issued simple directives such as "What is your name", "What is my name?", "How old are you?", What is your favorite color", etc. The directives became increasingly individualized for each participant, e.g. "What is your sister's name?" By the end of the sessions, the content had evolved to games with directives such as "Show me something that is X" where X is a color, "Find X" and "Look at X" where X is an object in office, and "Touch X" e.g. touch my elbow, now touch your elbow, now touch the SLP's elbow.

The sessions were video-recorded for subsequent analysis. We counted the number of times a subject responded correctly to the robot's directives (with SLP prompting) and the number of times there was no response. Table 1 shows results for Subject 2.

Table 1 Results for Subject 2 in the second study

	Total Directives	No Response	Correct Response
Sept 30 2014	32	13	19
Oct 6 2014	33	8	25
Oct 14 2014	50	2	48
Oct 28 2014	48	4	44
Nov 3 2014	45	6	39

By inspection, one can see that Subject 2's No Response count decreases and flattens out with a noticeable improvement in the 3rd session. Likewise, the Correct Response count consistently increases with noticeable improvement at the 3rd session. These counts are consistent with the increase of Total Directives over time. One would expect Total Directives to increase as No Response decreases because the SLP is spending less time prompting the subject.

The results for subject 3 are inconclusive. The subject performed well during the 1st session: Total Directives=16, No Response=1, and Correct Response=15. However, the subject was despondent during the 2nd session and was tantruming during the 3rd session, and thus was removed from the therapy in both cases. This results suggests that robot interaction may not be suitable for certain patients.

These results are mildly encouraging. The study issued quantitative results although it has limitations. The study gave us a sense of the problem of understanding the incremental benefit of co-robot therapy over well-defined participant groups, albeit with a small sample. Further, the methodology by which we evolved the games over time was useful. The main debate after each session was on the design or modification of robot behaviors for the next session, during which debate there was healthy tension between what was desired by the subjects' SLP and SPED teachers and what the robot programmer was able to implement; however, accommodations were reached.

4   The Third Study: Joint Attention, Turn-Taking, Initiative

The third study continued with Subject 2 at the same elementary school, starting in November of 2015 [9]. It is an 8-month study with the subject receiving 20-minutes sessions twice a week. The goal of the therapy is the same – to foster the subject's joint attention skills. This time the therapy sessions are conducted in the school's playroom. It is a large room with play objects scattered about, e.g. a swing, trampoline, blocks, toys, a yoga ball, etc. The configuration of the robot operator, the robot, the SLP, and Subject 2 is the same as the previous study. Fig. 4 provides context.

Fig. 4 Subject 2 in the third study playing the "What am I looking at?" game

Improvements in game content began early in the sessions. We introduced a "What am I looking at?" game in which the robot looks in alternative directions. Our measure of the subject's performance was refined by "Correct Response with No SLP prompting" and "Correct Response but Only with SLP Prompting." Further, correct responses were reinforced by verbal utterances from the SLP and/or the robot.

During the first ten cycles of the methodology, i.e. ten sessions over a five-week period, accommodations were made involving (i) the syntax and intonation of the robot's directives/reinforcements and game content and (ii) movement of the robot's head in the direction of interest. A new game "The Farmer in the Dell" was introduced in which the subject answers directives of the form "What does the X say?"

After the first few sessions, the SLP noted that the subject had made significant improvements: "He needs less prompting/encouragement to engage in an activity appropriately. He is initiating on a much more frequent basis (for him) and is not as difficult to redirect back to a task when he is distracted by internal thoughts.

He appears to attend to his peers more this year as well." It is not known the extent to which robot therapy contributed to the improvement. Other factors might include his music therapy, summer school activities, and natural biological maturation. Future research will address this question. In any case, an analysis of the first three sessions with Subject 2 in the third study is consistent with the SLP's statement, as shown in Table 2.

Table 2 Results for Subject 2 in the first three cycles of the third study

	Total Directives	No Response	Correct without Prompting	Correct with Prompting
Nov 2 2015	105	4	64	37
Nov 5 2015	95	7	60	28
Nov 9 2015	135	11	94	30

To date we have completed fourteen cycles of SSM. As an example: At the start of cycle 9 the problem situation had remained "improve skills in joint attention" but the subject had made considerable progress according to the SLP and SPED teacher (supporting the results in [12]). This led to considerations of related activities including the home (he likes to talk about the robot), routine school tasks (he gets bored with teachers but doesn't get bored with the robot), group therapy (no noticeable improvement because other group members have various deficits), and programming the robot (some behaviors are simple to implement but others are difficult and time-consuming). After debating these activities, it was agreed to find accommodations to introduce turn-taking and initiative skills.

Thus in cycle 10 the problem situation is restated "to improve the subject's skills in joint attention, turn-taking, and initiative." The games were improved as follows: During the "Everybody look at X" game, the robot asks the SLP and the subject to take a turn. If the subject responds, the robot turns its head in the direction of X. During the "What does the X say?" game, the robot likewise asks the SLP and the subject to take a turn. If the subject responds, the robot produces the sound of X. The SLP prompts the subject with cue cards "Everybody look at ___ " or "What did the ___ say?" During cycle 10 the robot asked the subject to take a turn 16 times. Video analysis showed that he responded correctly 11 times and responded correctly but only with SLP encouragement 5 times. Based on this result, it was agreed to continue these games several more cycles during which the SLP will try to fade out the cue cards.

5 Discussion and Future Work

The pilot studies have clear limitations. The sample is small. There is no control group, and there is no opportunity to study incremental benefits over diverse participant groups. However, the studies encourage us to continue this line of research, to compare improvements in joint attention, turn-taking, and initiative skills against changes in the system over the duration of the study, to assess improvements in these skills in external activities, and to develop a well-defined long-term experiment at the elementary school starting Fall/2016 with well-defined participant groups.

The robot is tele-operated. With the evolution of changes in the therapy protocol, the operator's console has become unwieldly. Fig. 5 shows the console on the 1st cycle and the 10th cycle. It gives a visual sense of the difficulty. The operator clicks on a box to execute an element of a game. Thus, the operator is busy clicking on boxes while a session unfolds. A better approach is to re-design the robot program with the agent perspective such that agents are modular, modifiable for future cycles, able to function independently, and robust. We also wish to render the robot partially autonomous, thereby reducing the need of a robot operator [7]. Thus, a new design will be based on the subsumption architecture [14]. Agents will reside on levels such that higher levels subsume lower levels. If higher-level agents become dysfunctional, the lower-level agents can continue to function albeit with less functionality than the system as a whole. We envision (i) an introduction agent, (ii) a multiplicity of game agents, (iii) a reinforcement agent, and (iv) a multiplicity of auxiliary agents, e.g. the boxes demarcated in red in Fig. 5 (right) control movement of the robot's head and are used in several games. As such, it will be an auxiliary agent called upon by game agents. A re-implementation using the agent perspective is scheduled for Fall/2016.

Fig. 5 Robot operator's console on the 1st cycle (left) and the 10th cycle (right)

References

1. US Center for Disease Control/Prevention. www.cdc.gov/ncbddd/autism/data.html (retrieved January 22, 2016)
2. US Department of Health and Human Services, National Institute of Health. www.nidcd.nih.gov/health/voice/pages/specific-language-impairment.aspx (retrieved January 22, 2016)
3. Kim, E., Berkovits, L., Bernier, E., Leyzberg, D., Shic, F., Paul, R., Scassellati, B.: Social Robots as Embedded Reinforcers of Social Behavior in Children with Autism. Journal of Autism and Developmental Disorders 43(5), May 2013
4. Scassellati, B., Admoni, H., Matarić, M.: Robots for use in autism research. Annual Review of Biomedical Engineering 14 (2012)

5. Diehl, J., Schmitt, L., Villano, M., Crowell, C.: The Clinical Use of Robots for Individuals with Autism Spectrum Disorders: A Critical Review. Research in Autism Spectrum Disorders **6**(1), January–March 2012. Elsevier

6. Diehl, J., Crowell, C., Villano, M., Wier, K., Tang, K., Riek, L.: Clinical applications of robots in autism spectrum disorder diagnosis and treatment. In: Comprehensive Guide to Autism. Springer, New York (2014)

7. Pennisi, P., Tonacci, A., Tartarisco, G., Billeci, L., Ruta, L., Gangemi, S., Pioggia, G.: Autism and social robotics: A systematic review. Autism Research, Wiley On-Line Library (2015)

8. Kim, E., Paul, R., Shic, F., Scassellati, B.: Bridging the Research Gap: Making HRI Useful to Individuals with Autism. Journal of Human-Robot Interaction **1**(1) (2012)

9. Lewis, L., Charron, N., Clamp, C., Craig, M.: Soft systems methodology as a tool to aid a pilot study in robot-assisted therapy. In: Late-breaking report, 11th International Conference on Human-Robot Interaction, Christchurch, New Zealand (2016)

10. Checkland, P., Poulter, J.: Learning For Action: A Short Definitive Account of Soft Systems Methodology, and its use for Practitioners, Teachers and Students. Wiley (2007)

11. Checkland, P.: Soft Systems Methodology: A Thirty Year Retrospective. Systems Research and Behavioral Science **17**(1) (2000)

12. Warren, Z., Zheng, Z., Swanson, A., Bekele, E., Zhang, L., Crittendon, J., Weitlauf, A., Sarkar, N.: Can Robotic Interaction Improve Joint Attention Skills? Journal of Autism and Developmental Disorders **45**(11), November 2015

13. NAO   Specification.   www.doc.aldebaran.com/2-1/family/nao_h25/index_h25.html (retrieved February 28, 2016)

14. Brooks, R.: A Robust Layered Control System for a Mobile Robot. IEEE Journal of Robotics and Automation **2**(1) (1986)

Towards an Integrated Approach to Diagnosis, Assessment and Treatment in Autism Spectrum Disorders via a Gamified TEL System

Laura Tarantino, Monica Mazza, Marco Valenti and Giovanni De Gasperis

Abstract Autism Spectrum Disorders (ASDs) are characterized by atypical patterns of behaviors and impairments in social communication and interactions. Information and Communication Technologies (ICT) have been recognized to have great potential in supporting ASD treatment: ICT-based tools are enjoyed since interaction with computers supports imagination of behaviors necessary for role-play in predictable environments. Differently from most proposals in the literature focused on treatment from the patient side, we are here interested in discussing how moving from paper-and-pencil measures to ICT tools may help psychologists and therapists in their diagnosis activities as well as in conceiving novel technology-enhanced interventions. In particular, we will present some features of a system under development aimed at pursuing an integrated approach including diagnosis and gamified learning.

Keywords Autism Spectrum Disorder · TEL · Gamification

1 Introduction

Autism Spectrum Disorders (ASDs) are characterized by restricted, repetitive and stereotyped behavior and core deficits in social communication and interaction [1].

L. Tarantino(✉) · G. De Gasperis
Department of Information Engineering, Computer Science and Mathematics,
University of L'Aquila, Via Vetoio 1, 67100 L'Aquila, Italy
e-mail: {laura.tarantino,giovanni.degasperis}@univaq.it

M. Mazza · M. Valenti
Department of Applied Clinical Sciences and Biotechnology,
University of L'Aquila, Via Vetoio 1, 67100 L'Aquila, Italy
e-mail: monica.mazza@cc.univaq.it, marco.valenti@univaq.it

M. Valenti
Centro di Riferimento Regionale per l'Autismo, ASL 1 Regione Abruzzo,
Via Lorenzo Natali 1, loc. Coppito, 67100 L'Aquila, Italy

© Springer International Publishing Switzerland 2016   141
M. Caporuscio et al. (eds.), *mis4TEL*,
Advances in Intelligent Systems and Computing 478,
DOI: 10.1007/978-3-319-40165-2_15

Social impairments severely interfere with the process of building relationships, functioning occupationally, and integrating and participating into community [2]. The importance of accurately identifying individuals with ASDs, and therapeutically intervening on them, has never been greater: in the last decade the prevalence of autism has risen dramatically, with a systematic review estimating global prevalence of pervasive developmental disorders at 62 per 10,000 and autistic disorder at 17 per 10,000 [3] (with figures even higher in more developed countries [4]).

If no definitive reason for such an increase has been determined yet, it is certain that more children are identified as having ASD than even before, that these children need help, and that school systems and medical care systems require increasing resources to ensure that people diagnosed with ASD achieve optimal outcomes and improve their quality of life. Accessibility of treatment is to be considered a key factor: traditional behavioral treatment requiring intensive sessions, e.g., are not accessible to a vast majority of individuals with ASD due to intervention costs.

The causes of autism remain largely unknown, though there is evidence that genetic, neurodevelopmental, and environmental factors are involved. Deficit in social cognition and its components such as Theory of Mind (i.e., the ability to understand another's thoughts, beliefs, and other internal states), emotional contagion, and prosocial behavior were reported in persons with ASD [5,6]. Given the influence that, due to studies on the brain areas related to social cognition abilities [7], the social cognition model in ASD has in research and practice, it would be crucial to develop evidence-based intervention strategies based on social cognition skills.

To this aim, Information and Communication Technology (ICT) has great potential: ICT-based tools – typically enjoyed by individual with ASD since interaction with computer does not pose sever expectations and judgement issues and allows to discover conventions in a safe and predictable environment – can support the imagination of contexts, people and behaviors necessary for role-play, and offers replicability, possibility of modifiable multisensory stimulation, and capacity to implement individualized intervention (see, e.g., [8,9]). The literature generally agrees on the potential benefits of technology-enhanced approaches in supporting the learning process of emotional and mentalizing competencies in individuals with ASD. Anyhow it has to be said that, despite promising preliminary results, studies of ICT-based tools for the treatment of ASD have to be considered still at their infancy because of limitations on the complexity of the proposed tools (typically built ad hoc for a specific study) and on the size of the clinical groups involved in the evaluation (for example, out of the 38 ICT-based studies surveyed in [10], 20 involved a sample with a size below 10 or not even specified and only 3 have been evaluated with more than 30 persons), hence suffering, in summary, from the lack of real generalizable results and scholarly knowledge beyond the context of proof of concept studies.

We argue that it is necessary to enlarge the focus of technological interventions, aiming at systems able to support therapists in the multifaceted objective of creating personalized learning material, customizing interventions, and monitoring/assessing progresses (possibly in a context of tele-rehabilitation). This would allow researchers to conduct longitudinal studies and to evaluate whether improvements reached in the interaction with a tool actually hold in real life situations.

In this paper we report on first steps in this direction, carried out under a participatory design approach by a multidisciplinary team, including psychologists, computer scientists, and persons with ASDs, aimed at developing *an integrated suite of tools for diagnosis, gamified technology-enhanced learning, and assessment.* In particular, in Section 2 we single out key features to be addressed when conceiving ICT-based tools for ASD treatment and sketch our vision to overcome limitations of the field. Then, in Section 3 and Section 4 we discuss a multidisciplinary study based on such a vision: Section 3 sketches the organizational situation to ameliorate, in terms of diagnostic/assessment measures currently used by therapists, while Section 4 discusses how moving from a paper-and-pencil to an ICT-based approach may help psychologists and therapists achieving an integrated technology-enhanced framework for diagnosis, treatment, and assessment. In Section 5, conclusions are drawn.

2   ICT and ASD: A Review of Strenghtness and Limits

In the last decade, research has explored a variety of ICT-based approaches to ASD with diverse goals: (i) as assistive technologies, to counteract the impact of sensory and cognitive impairments on daily life; (ii) as rehabilitation tools, to modify and improve the core deficit in social cognition, and (iii) as education tool, to help children acquire social and academic skills [11]. Initial fear that the use of technology might further isolate ASD sufferers who have problems in social relationships [12] have been overcome by evidences about success of ICTs to improve social interaction when used correctly [13]. While for extensive reviews we refer, e.g., to [10,11,14,15], here we want to single out key features to be considered when designing an ASD-oriented technology-enhanced treatment and discuss limitations to overcome.

Learning Benefits. A decision to be taken when designing a technology-enhanced tool able to teach aspects of social functioning to people with ASDs is about the social skills to be targeted. As underlined in [8], the skills chosen should be relevant to many different situations thus offering the opportunity to investigate generalization of skills between contexts. To date, studies related to technology-enhanced interventions have addressed a variety of social skills that are regarded essential for persons with autism, including the ability to recognize faces and emotions; to initiate, maintain, or terminate a behavior; to improve vocabulary and reading skill, spatial planning, functional activities of daily living; to enhance vocal imitation [15].

Sensory Channels. Multimediality, and the possibility of modifiable multisensory stimulation, is one of the main strenghts of technology-enhanced treatment. Computer graphic displays make abundant use of visually cued instructions, which are recommended for interventions in ASD [8] (e.g., baloons may be used as visual aids to suggest what characters in a scene are thinking or feeling).

Use of Avatars. Conversational avatars playing the role of an instructor are proven to advance the educational process [17] and to improve social skills [18]. Emotionally expressive avatars appears to be crucial [19], even better if they have voices [17].

Virtual Environments (VE). Computer-generated representations of environments with realistic appearance appear promising for teaching social understanding, due to their capability of illustrating scenarios representing situations that may not be feasible in therapeutic settings, of promoting role-play, and of allowing participants to take a first-person role for skill-learning in virtual social situations [8,20,21]. Studies report that, by reproducing a "virtual cafè" to teach social skills, the speed of execution of social tasks in the VEs improved after the repetition of the task and learned skills were transferred into a VE proposing a situation with similar demands (on the bus) [8]. It is argued that the realism of the simulated environment increases the probability that the person with ASD transfers learned skills into everyday live ([8,22]), though longitudinal studies that support such an hypothesis are missing.

Gamification. Interaction with VEs is generally perceived as playful and it is proved that subjects learn while they play [23]. Not surprisingly, a great number of pilot studies on VEs are based on serious games (e.g., [8,9,24]). The gamified approach to learning requires to design situations at different levels of social complexity and to define criteria for progressing in the game. While most studies are designed to chain learning based only on performance aspects (i.e., on correctness of the answers), some proof-of-concept studies proved that an adaptive approach including also criteria based on physiological markers of engagement (e.g., pupil dilation and blink rate) contributes to improved performance of participants, while leaving the generalization of skill improvements in real-life an open question [9].

Devices. Though traditionally technology-enhanced treatment has been based on desktop computers with mouse and possibly joystick [20], most of the recent research projects use touch screens [11, 25]. Authors in [26] conducted a systematic review of studies involving iPods, iPads, and related devices in teaching programs for individuals with developmental disabilities. The 15 studies were largely positive and showed that these devices are viable technological aids for such individuals. These results appears important since they open the possibility of accessible low-cost technology enhanced treatments possibly in a tele-rehabilitation context.

Limits. As underlined by [15], recent literature tends to emphasize the potential of technology more than its demonstrated effectiveness: technology-based treatment is still perceived as "emerging" rather than "established", and the clinical validity is still a matter of debate, the main flaws being lack of statistical significance in the clinical group they are based on, limited complexity of the tools, designed ad-hoc for the pilot studies and not supporting creation of learning material, and lack of longitudinal studies proving the transfer of acquired skills from ICT-based tools to daily life.

Our Vision. We argue that, in order to move from potentiality to proven effectiveness, it is necessary to take one step backward and put new basis for more systematic research. This requires, first of all, to analyze the work of researchers, therapists, and operators in ASD diagnosis, assessment, and treatment, and conceive technology-enhanced approaches that allow them to pursue the multiple objectives of (i) improving the organization of their work, (ii) testing and comparing in a systematic way the diverse options that computers offers, e.g. with respect to (multi)sensory channels and physiological markers of engagement (e.g., by eye trackers), (iii) applying technology-based treatments, and (iv) evaluating them in longitudinal studies. Furthermore, a multiplatform approach including also low cost ICT devices, would allow to reach – through tele-rehabilitation - individuals with ASD that, for various reasons, remain otherwise marginalized and not treated.

3   The Organizational Situation

To address these goals we launched a multidisciplinary project based on the cooperation between the Department of Information Engineering, Computer Science and Mathematics, the Department of Applied Clinical Sciences and Biotechnology, and the Center for Autism of the University of L'Aquila, and the Abruzzo Regional Reference Center for Autism. The team includes three psychologists, one medical doctor, three computer scientists, five young persons with ASD in the age range 15-23 involved according to a participatory design approach, and their families involved as informant. According to our vision, the first phase of the project has been focused on a collaborative analysis of the organizational situation, e.g., of diagnosis and assessment practices, to single out possible weakness to ameliorate. This study was conducted through semi-structured interviews with therapists, focus groups with persons with ASD and their families, and field studies.

The Organizational Situation. Following state-of-the-art approaches, individuals that arrive to the ASD centers undergo a number of standardizes clinical measures to possibly diagnose ASD (e.g., the Autistic Diagnostic Observation Schedule-2 (ADOS-2) [27] and the Autism Diagnostic Interview-Revised (ADI-R) [28]). Once diagnosed as having ASD, individuals undergo a number of cognitive and social cognitive measures, administered as paper-and-pencil tasks (Table 1 sketches some of them to give an idea of aims, structures, content, and demands on the participants).

Table 1 Paper-and-pencil social cognition measures, typically administered in random order.

Main social cognition tasks (summary)
Emotion Attribution Task [29]. It assesses the ability to represent the emotions of others. The participant is presented with 58 short stories (typically no longer that one or two sentences) describing an emotional situation (e.g., "Charles is lying in the forest. A poisonous spider falls on his chest") and is required to provide an emotion (by writing it down) describing how the main character might feel in that situation.
Advanced Theory of Mind Task [29,31]. It consists of 13 very short stories of different type (e.g., Lie, Joke, Pretend) accompanied by two questions regarding comprehension (e.g., "Is it true what the character said?") and justification (e.g., "Why did he say that?").
Social Situation Task [29]. It consists of 25 medium size stories including a number (1 to 3) of social situations that the participant is asked to judge as "normal", "a bit weird", "quite weird", or "extremely weird".
Attribution of intentions [32]. It consists of fifty 3-picture comic strips illustrating everyday social scenarios. After each strip the participants is presented with three pictures illustrating alternative endings of the story and is asked to select the one that best complete the story from a logical point of view.
Eyes task [30]. The participant is given 36 photographs depicting the ocular area in an equal number of actors and actresses. At each corner of each photo, four mental state descriptors are printed, only one of which correctly identifies the depicted person's mental state.

Main Results of the Collaborative Analysis. A primary problem emerged from the analysis: while structure, content, and administration rules of the tasks would clearly allow a straightforward digitalization, no computer-based support is actually available, and operators have the burden of integrating results from clinical, cognitive, and social cognitive measures, in order to record them, analyze them, and plan treatment. A second type of problems is related to the relationship between the administering operator and the administered person, with the latter frequently showing discomfort for the presence of the former (e.g., in tasks including stories (particularly when of medium size) participants prefer to read the story by themselves and want to feel free to re-read (parts of) it before answering the questions). Other issues are related to long-term relationships with treated kids, with therapists reporting the difficulties, in some cases, to maintain a regular contact with them, e.g., for logistic and/or economic reasons.

4   From Paper-and-Pencil to Technology Enhancement

Starting from the results of the analysis, we planned a roadmap of interventions (not necessarily strictly sequential) aimed at modifying the organizational situation by introducing a multi-user ICT-based system conceived as a coherent suite of tools, envisioning an integrated diagnosis/assessment/treatment approach (see Table 2). As to the treatment, we are working under the theoretical premise that social skills may be learned through a suite of serious games addressing diverse deficits and conceived as gamified variants of some of the tasks included in the assessment strategy (in particular, all tasks included in Table 1 are gamifiable), within the framework of an adaptive Technology-Enhanced Learning (TEL) approach.

Table 2 Areas of interventions for the development of the integrated suite.

The ICT-based interventions
Digitalization – Implementing digital (interactive) versions of the tests as is (basic step).
Sensory enrichment – Exploiting multimediality for exploring sensory alternatives (verbal, pictorial 2D, pictorial 3D, interactive pictorial 3D, audio, possibly in combination) for the tasks administration, to evaluate their differences in terms of contribution to the assessment and precision (e.g., the short stories of the "Emotion attribution task" may be rendered purely verbally, with or without audio, or by relying on dual coding (verbal plus pictorial), with or without audio, in 2D or 3D, with or without an instructional avatar).
Gamification – Translating "gamifiable" tests into serious games in order to achieve a coherent suite of games specialized for different skills. The playing environment includes a customizable conversational avatar guiding participants during the playing of the games, able to interact with objects and characters of the scenes, and providing individualized feedback based on participants' responses to contribute to the acquisition of social skills (e.g. "your friend would have felt more comfortable if you had paid more attention on how he is feeling. Try again and make him comfortable").
Use of interactive patterns – Exploiting visual patterns for supporting participants during tasks/games activities (e.g., stories of the "Social Situation Task" are broken down into episodes and visually represented by an interactive carousel thus allowing participants to freely browse them)
Multiple platforms/settings – Developing two versions of the games: in the full version adaptivity is based on both performances and physiological markers of engagement (requiring equipment available only in lab setting), while a light version relies on a performance-based adaptive mechanism and can run on accessible low-cost devices (e.g., tablet) thus allowing persons with ASD to continue the treatment at home. Operators may monitor players' activities and progresses.
Creation of learning material – Supporting experts in the creation and refinement of (learning) material.

The system is being implemented according to a multitier architecture able to decouple content and rendering. The definition of 3D scenes is based on the open-source WebGL library (http://www.webgl.org) a cross-platform, royalty-free web standard for a low-level 3D graphics API, based on OpenGL ES 2.0.

5   Conclusions

In this paper we discussed issues behind a radical change of perspective in the studies on technology-enhanced treatment for persons with ASD. Literature agrees on the potential benefits of technology-enhanced approaches to the acquisition of social skills, but so far studies have been focused on proof-of-concept prototypes and evaluated with limited clinical groups. We argue that, to move from "emerging" to "established" treatments with proven effectiveness, it is necessary to take one step backward to put new basis for more systematic research starting from the analysis of the work of therapists and operators to conceive systems that, exploiting results that the field has achieved so far, support longitudinal studies on TEL systems for people with ASD. In the paper we sketched our vision and an ongoing project aimed at realizing an integrated suite of ICT-based tools able to support both operators in diagnosis/assessment activities and people with ASD in their treatment by a gamified adaptive TEL system. Particular attention is being put on accessibility of the tools to be able to deliver treatment regardless of barriers of distance, time, and cost.

References

1. American Psychiatric Association: Diagnostic and Statistical Manual of Mental Disorders: DSM-V. 5th edn. American Psychiatric Publishing, Arlington, VA (2013)
2. Fletcher-Watson, S., McConnell, F., Manola, E., McConachie, H.: Interventions based on the Theory of Mind cognitive model for autism spectrum disorder (ASD). Cochrane Database Syst. Rev. **3:CD008785** (2014)
3. Elsabbagh, M., Divan, G., Koh, Y.-J., Kim, Y.S., et al.: Global prevalence of autism and other pervasive developmental disorders. Autism Research **5**, 160–179 (2012)
4. Baird, G., Simonoff, E., Pickles, A., Chandler, S., Loucas, T., Meldrum, D., et al.: Prevalence of disorders of the autism spectrum in a population cohort of children in South Thames: the Special Needs and Autism Project (SNAP). Lancet **368**, 210–215 (2006)
5. Zaki, J., Ochsner, K.N.: The neuroscience of empathy: progress, pitfalls and promise. Nat. Neurosci. (2012)
6. Lai, M.C., Lombardo, M.V., Baron-Cohen, S.: Autism. Lancet (2014)
7. Dziobek, I., Preibler, S., Grozdanovic, Z., Heuser, I., Heekeren, H.R., Roepke, S.: Neuronal correlates of altered empathy and social cognition in borderline personality disorder. NeuroImage **57**, 539–548 (2011)
8. Mitchell, P., Parsons, S., Leonard, A.: Using Virtual Environments for Teaching Social Understanding to 6 Adolescents with Autistic Spectrum Disorders. J. Autism Dev. Disord. **37**, 589–600 (2007)
9. Lahiri, U., Bekele, E., Dohrmann, E., Warren, Z., Sarkar, N.: A Physiologically Informed Virtual Reality Based Social Communication System for Individuals with Autism. J. Autism Dev. Disord. **45**, 919–931 (2015)
10. Aresti-Bartolome, N., Garcia-Zapirain, B.: Technologies as Support Tools for Persons with Autistic Spectrum Disorder: A Systematic Review. Int. J. Environmental Research and Public Health (2014)
11. Boucenna, S., Narzisi, A., Tilmont, E., et al.: Interactive technologies for autistic children: A review. Cognitive Computation **6**, 722–740 (2014)
12. Powell, S.: The use of computer in teaching people with autism. In: Autism on the Agenda: Papers from a National Autistic Society Conference (NAS 1996), London, England (1996)
13. Ploog, B.O., Scharf, A., Nelson, D., Brooks, P.J.: Use of computer-assisted technologies (CAT) to enhance social, communicative, and language development in children with autism spectrum disorders. J. Autism Dev. Disord. **43**, 301–322 (2013)
14. Grynszpan, O., Weiss, P.L., Perez-Diaz, F., Gal, E.: Innovative technology-based interventions for autism spectrum disorders: A meta-analysis. Autism **18**, 346–361 (2013)

15. Parsons, S., Cobb, S.: State-of-the art of virtual reality technologies for children on the autism spectrum. European Journal of Special Needs Education **26**, 355–366 (2011)
16. Quill, K.A.: Instructional considerations for young children with autism: the rationale for visually cued instruction. J. Autism Dev. Disord. **27**, 697–714 (1997)
17. Konstantinidis, E.I., Luneski, A., Frantzidis, C.A., Pappas, C., Bamidis, P.D.: A proposed framework of an interactive semi-virtual environment for enhanced education of children with autism spectrum disorders. In: The 22nd IEEE International Symposium on Computer-Based Medical Systems (CBMS) (2009)
18. Hopkins, I.M., Gower, M.W., Perez, T.A., Smith, D.S., Amthor, F.R., Wimsatt, F.C., Biasini, F.J.: Avatar assistant: improving social skills in students with an ASD through a computer-based intervention. J. Autism Dev. Disord. **41**, 1543–1555 (2011)
19. Fabri, M., Awad Elzouki, S.Y., Moore, D.: Emotionally expressive avatars for chatting, learning and therapeutic intervention, human-computer interaction. In: HCI Intelligent Multimodal Interaction Environments, pp. 275–285 (2007)
20. Bellani, M., Fornasari, L., Chittaro, L., Brambilla, P.: Virtual reality in autism: state of the art. Epidemiol. Psychiatr. Sci. **20**, 235–238 (2011)
21. Parsons, S., Mitchell, P., Leonard, A.: The use and understanding of virtual environments by adolescents with autistic spectrum disorders. J. Autism Dev. Disord. **34**, 449–466 (2004)
22. McComas, J., Pivik, J., Laflamme, M.: Current uses of virtual reality for children with disabilities. Studies in Health Technology and Informatics **58**, 161–169 (1998)
23. Vera, L., Campos, R., Herrera, G., Romero, C.: Computer graphics applications in the education process of people with learning difficulties. Comp. & Graphic. **31**, 649–658 (2007)
24. Serret, S., et al.: Facing the challenge of teaching emotions to individuals with low- and high-functioning autism using a new Serious game: a pilot study. Molecular Autism **5**, 37 (2014)
25. Konstantinidis, E.I., Bamidis, P.D., Koufogiannis, D.: Development of a generic and flexible human body wireless sensor network. In: Proceedings of the 6th European Symposium on Biomedical Engineering (ESBME) (2008)
26. Kagohara, D., et al.: Using ipods and ipads in teaching programs for individuals with developmental disabilities: A systematic review. Res. in Developmental Disabilities **34**, 147–156 (2013)
27. Lord, C., Rutter, M., Di Lavore, P.C., Risi, S., Gotham, K., Bishop, S.L.: Autism Diagnostic Observation Schedule (ADOS-2): Manual, 2nd edn. Western Psychological Services, Los Angeles (2012)
28. Rutter, M., Le Couteur, A., Lord, C.: Autism diagnostic interview revised WPS edition manual. Western Psychological Services, Los Angeles (2003)
29. Blair, R.J., Cipolotti, L.: Impaired social response reversal. A case of "acquired sociopathy". Brain **123**, 1122–1141 (2000)
30. Baron-Cohen, S., Wheelwright, S., Hill, J., Raste, Y., Plumb, I.: The "reading the mind in the eyes" test revised version: a study with normal adults, and adults with Asperger syndrome or high-functioning autism. J. Child Psychol. Psychiatry **42**, 241–251 (2001)
31. Happè, F.G.E.: An Advanced Test of Theory of Mind: Understanding of Story Characters' Thoughts and Feelings by Able Autistic, Mentally Handicapped, and Normal Children and Adults. J. Autism Dev. Disord. **24**, 129–154 (1994)
32. Sarfati, Y., Hardy-Bayld, M.C., Besche, C., Widlocher, D.: Attribution of intentions to others in people with schizophrenia: a non-verbal exploration with comic strips. Schizophrenia Research **25**, 199–209 (1997)

Feedbacks in Asynchronous Activities in Virtual Learning Environments: A Case Study

Gerlane Romão Fonseca Perrier and Ricardo Azambuja Silveira

Abstract This paper presents some results of a case study, where we intend to identify the role of tutors in communicative processes, and their contribution to improve the meaningful learning through an accurate analysis of the content posted in discussion forums tools in a Virtual Learning Environment, used for communication and feedback to the students. The research analyzed the participation of teachers and tutors in six disciplines of a distance learning technical course in food agricultural college. As result of the research, it concludes that the discussion forum is very important to improve experience interchange, significant reflection and social construction of knowledge.

Keywords Feedback · Mentoring · Discussion forum · Asynchronous communication

1 Introduction

This article aims to discuss the role of the tutor as facilitator agent of meaningful learning by using asynchronous communication tools in Virtual Learning Environments. It contains results from the Course Work Completion to the Specialization in Management and Teaching in Distance Education by Perrier [1].

We identify the tutor's responsibilities in conducting the teaching-learning process in distance education and the importance of interactivity to the success of the learning process in distance.

G.R.F. Perrier
CODAI/UFRPE, Universidade Federal Rural de Pernambuco,
Rua Dom Manoel de Medeiros, s/n, Recife, PE, Brazil
e-mail: gerlaneperrier@gmail.com

R.A. Silveira(✉)
PPGCC/UFSC, Universidade Federal de Santa Catarina,
Campus Reitor João David Ferreira Lima, Florianópolis, Santa Catarina, Brazil
e-mail: ricardo.silveira@ufsc.br

© Springer International Publishing Switzerland 2016
M. Caporuscio et al. (eds.), *mis4TEL*,
Advances in Intelligent Systems and Computing 478,
DOI: 10.1007/978-3-319-40165-2_16

In the educational process, communication is essential for exchanging of knowledge and, thus, promotes the construction of learning. In distance education, physical and temporal distances are the key features that set it apart from regular education. In this context, the communication among faculty and students deserve special attention by those involved in the process, whether teacher educators, tutors or students. Tutor is mainly responsible for conducting the interactive process through the discussion forums. However, we intend that there is a gap in theoretical references about possible types of tutorial performance and contributes to improve meaningful learning.

This article presents the results of our research that seeks to investigate how important is the feedback in asynchronous learning activities performed in virtual learning environments by deeply examining the discussion forum among students and teachers in real life classes. The next section presents the theoretical background of the research, the third section describes the methodology, the forth one presents some findings, the fifth section discusses these results and the last one present some conclusions.

2   Background

There are several pedagogical models to explain the construction of learning processes. The social-historical theories of Vygotsky and Feuerstein's Mediated Learning, and the concept of inter-cognition proposed by Bicalho [2] and Bicalho and Oliveira [3] base this research.

According to the Theory of Meaningful Learning, as proposed by Ausubel [4], the school should provide a learning environment that takes into account the cognitive, emotional and social reality of the student or the group of students, providing situations that allow significant learning of certain skills.

The tutor should take the pedagogical commitment to promote the internalization of knowledge through its interactive actions, taking into account the realities of the students in order to encourage them to reflect on their previous knowledge and thereby ensure meaningful learning. To do this, we should use interactivity strongly as an important tool for learning, and encourage the exchange of knowledge and experiences, which it may be internalized in the learner's mind.

According to Bicalho and Oliveira [3], inter-cognition is the result of collective and dialogical exchange of knowledge through interpersonal communication, and occurs when the process of internalization of knowledge by reframing acquired among the dialogic processes in a continuous process of co-authoring.

According to Ausubel [4], information given in feedback interacts with prior knowledge, promoting learning. Through the feedback, participants are conscious of how they should behave, interact, say, think and do something in a particular environment in order to accomplish their goals. The feedback can enhance behavior and thereby stimulate learning, as a powerful tool to guide the learning and stimulate the student to think about their responses or actions, encouraging him with purification and hence individual development.

Feedback is a type of information transmission in the educational process. There are several different types of feedback:

1. Recognition for success in response of an action, considered as positive feedback;
2. Incentive to encourage the improvement of some task that, while not wrong, is incomplete, or that deserves further consideration. This is considered a constructive feedback;
3. Warning about the quality of response or action, considered unsatisfactory by those offering feedback, this is the negative feedback.

Williams [4] presents some basic principles that is needed to consider for improving the feedback, given the importance of language and personal interaction in a virtual learning environment:

4. Quality: the quality and quantity of feedback offered are important tools for valuation of personal and interpersonal relationship;
5. Clarity: feedback should be clear enough for that message can hits the target;
6. Dialogue: In the distance education context the quality of the dialogue produced by interactive feedback can contribute to the approach or spacing the actors;
7. Cordiality: Mutual respect is essential for the dialogue between the actors. Thus, the feedback should always be cordial;
8. Eye contact: essential in the face communication in distance education, it is not always possible to have synchronous communication event;
9. Opportunity: the absence of communication generates real trouble with serious consequences for those who looks forward feedback (deny feedback is like giving a psychological punishment).

According to Abreu e Lima and Alves [6], the feedback ladder model offers support to students thinking, aiming to establish a culture of trust and constructive support among teachers and students and among the students. This feedback model develops in four stages: to clarify, to resolve possible doubts and better lead the discussions to achieve the proposed objectives; to appreciate any and all contributions, to recognize the importance of participation; to question, by asking questions that encourage reflection and thus contribute to the clearance of previous answers; and to suggest new research and reflections, to prevent that the student could be satisfied with just the first answer and be challenged to strive for meaningful learning.

In the sandwich feedback model [6], the comment about inappropriate behavior comes from two actions or positive behaviors, comprising three steps: highlight, in their responses, something positive; suggest improvements, demonstrating that it is always possible refine an answer; and close focusing in the best of a response.

3   Methodology

We conducted the research observing the participation of former teachers and
tutors of a Technical Course in Food at the Agricultural College in six different
classes and subject matters.

Fifty vacancies were available in each distance education pole. Each pole was
assisted by a tutor, designed to follow the learning process in each discipline. We
therefore analyzed the participation of fourteen tutors and seven professors in
asynchronous activities in this disciplines, with emphasis on how they are
contributing (or not) with the learning process.

This review took place after the 150MB of data recovery with compressed
backups of disciplines, kindly provided by the General Coordination of Distance
Education Center/CODAI[1].

The content of the messages posted in the discussion forums were analyzed by
focusing the interactions between teachers, tutors and students and the inter-
cognitive processes that contribute to the construction of meaningful learning.

The analysis of the feedback provided by the tutors in each discussion forums,
aimed to assess their impact on the learning process. It sought to then identify and
characterize, in equity, the different stages of feedback models to be analyzed and
how these can be evaluated for their contribution to the meaningful learning
process, according to the models' feedback ladder "and feedback sandwich,
presented earlier.

In the forums messages posted by subjects were selected some interactive
sequences in which it was possible to identify the dimensions of the internal
dynamics of the dialogue, which follows the relevant and irrelevant direction of
inter-cognition and meta-communicative processes described by Bicalho [2].
We followed six disciplines:

- Methodologies for Distance Learning: 60 hours
- Instrumental Portuguese: 80 hours
- Health and Safety at Work: 80 hours
- Hygiene in the Food Industry: 80 hours
- Basic Microbiology: 80 hours
- Biochemistry: 80 hours

4   Results

In order to demonstrate the importance of participation of the tutor in the conduct
of meaningful learning process, we selected as illustrative examples some
fragments collected from the interactive sequences posted in discussion forums of
the analyzed subjects of the research:

---

[1] CODAI – Colégio Agrícola Dom Agostinho Ikas, agricultural technical college supported
by Federal University of Pernambuco, Brazil.

— **Example 1:** We select an extracted example of the contribution of a tutor feedback as answer to the topic under discussion.

> *"It would be good that all people have the notion of first aid. In companies that have the technical SESMT the nursing technician or other health professional can do this training with the employees"*

In this feedback, the contribution of the tutor closes an idea or personal thought but not instigates discussion and does not prevent any student could instigate the development of new discussions on the topic, however it is expected to be role of mentoring drives this process.

— **Example 2:** We excerpted this example from a discipline where it fell tutors conducting the proposed discussions in the forums:

> *"Good night dear students. I see that both (name a student and name b student) studied the same source of linguistic variation. Would not be interesting that now you stood with their words, their understanding of this evolution of the Portuguese language?"*

In this feedback tutor sought to stimulate reflection for the construction of learning through the pursuit of personal opinions, without disregarding the research conducted by the students to try to present an answer to the question asked.

— **Example 3:** In this example in particular, we observed effective work performed by the tutors through its interventions in the conduct of the interactive process, according the feedback ladder model.

Initial response of a question about the understanding of students in relation to the biochemistry and the importance of their knowledge in the technical course in food:

> *"In my view Biochemistry is very present in our daily lives, for we are living beings, and we feed, then bio-molecules act in our body, but do not think their biochemistry in all things, a few minutes broke a stone ceramic, I think it was not amino acids, proteins, carbohydrates etc."*

First feedback provided by the tutor:

> *"Good post, (Name of a student). However, it is the only biochemical studies the proteins, carbohydrates etc. Does the composition of the stone contain a mineral? Will we need minerals too? Do we search? I suggest a reading of minerals and then return to the teacher's question."*

Response to the first feedback:

> *"I will search for minerals yes, because I think the stone has to do with them."*

Feedback response:

> *"Yes, (Name of a student). Note the expansion of your knowledge."*

Reflected Response of the student:

"I researched a bit and found that the bio-molecular reactions are present in all or almost all, our life is a real biochemistry, we live surrounded by things that undergo reactions at all times."

Second feedback:

"Good post, (Name of a student). What about the minerals, what were your findings in this matter?"

Response to second feedback:

"I did some research on minerals from our food are critical in the metabolic processes of the organism, each micro-nutrient plays an important role and in many cases interact to perform joint functions such as antioxidants, better calcification of bones, vitality, energy etc."

Third feedback:

"Excellent research, (Name of a student). Now he noted the importance of minerals. At the beginning of our discussion, you posted about breaking a ceramic stone. Pottery is clay and mud made. For example, the clay containing aluminum, which is a chemical element, as well as calcium contained in bones. In such cases, (name of (a) student), we are not talking about biochemistry, but talking about chemistry. When we talk about biochemistry, we speak of these chemicals in a living organism. In other words, we have the chemistry at all, but the biochemical we have when we talk about organisms. How could I answer the question the teacher, now with this new knowledge?"

Response to third feedback:

"I understood that biochemistry is when it comes to living organisms; e.g. those animals etc., and the chemistry subtends to other organisms, e.g. Stone pottery that I mentioned at the beginning"

Feedback response:

"Yes, (name (a) student), excellent participation! If possible, continue to participate with us in our discussions, because your posts are very important for us."

We observed that tutor conducted the learning process through the feedback ladder (clarifies, values, questions, suggests) and that it allows the internalization of experiences and prior knowledge, articulating it and resulting in what we know as meaningful learning.

5   Critical Analysis

To evaluate the impact of interactive participation of tutors in asynchronous activities, we observed the content of the forums used in the evaluated subjects to identify what kind of interaction were more effective for the meaningful learning process.

While living with subjectivity, we admit, in accordance with Moreira [7], that the significant learning is the result of interaction between new knowledge and prior knowledge, in which the new knowledge acquires meaning for the learner and prior knowledge gets richer, more differentiated, more elaborate in terms of meaning, and get more stability. Thus, from the contributions of tutors and teachers, we assess if these have been shown to stimulate reflection and able to change the behavior of students promoting meaningful learning.

We also accompanied the responses of students in order to verify whether the interventions of tutors encouraged participation and reflection on the answers given, in order to ensure the continued evolution of the internalization of knowledge.

Table 1 Analysis of the performances of teachers and tutors in the discussion forums

Course	Teacher performance	Tutors performance
Methodologies for Distance Learning	Effective participation, ensuring the direction of the discussions to the goals set and encouraging continued reflection of participants.	Effective participation of tutors, keeping the direction of the discussions to the goals set and encouraging continued reflection of the participants.
Instrumental Portuguese	He participated in a single forum, practically limited to the opening of others.	Merely bureaucratic participation with little stimulus to the discussions. Priority to other activities (chat, sending files - tasks).
Health and Safety at Work	He did not participate, but simply to open forums.	Effective participation of both tutors, but with little encouragement to the discussions. Stimulated the reflections, but not the continuity of post-reflection discussions.
Hygiene in the Food Industry	Monitor discussions, intervened to the appropriate direction to achieve the desired result.	Restricted and bureaucratic involvement of both tutors, contributing little to the reflection and construction of individual or collective learning.
Basic Microbiology	He did not participate, but simply to open forums.	Effective participation of tutors, but with a predominance of just one, which proved to be better prepared to lead the discussion and stimulate reflection on the part of students.
Biochemistry	He did not participate, but simply to open fo-rums.	Effective participation of both tutors, encouraging reflection and expansion of discussions contributing effectively to the learning process of the students from the contents of the discussion.

The analysis of the forums confirms the observation of the importance of the role of tutors in the conduct of the educational process by encouraging reflection as a means of self-improvement and construction of meaningful learning.

The method of feedback to be used is not imposed as a single, however, the analysis of the feedback submitted, you can see that can somehow there is a tendency towards standardization of form in the various disciplines, perhaps as an indication or request the teacher trainer, but without this presents itself consciously or institutionalized form. Table 1 summarize how teachers and tutors conducted the forum discussions.

The most important aspects in the content of the feedback provided by tutors who can stand out and which corroborates the observations are:

- Encouraging reflection is the main attribute of the feedbacks of the responses provided in the discussion forums.
- The warmth and clarity in texts produced are essential to avoid them called "communication noise" that comes to the problems arising from difficulties that may have the messages receivers decode them.
- Regardless of the content under discussion, the tutor must take the conduction of the educational process transmitting the useful and necessary concepts and values for self-learning and collaborative learning.
- Most of the tutors used the sandwich feedback model, even if they have internalized in their performances the model steps as such.
- The feedback ladder model can be clearly observed in the discussion forums of Biochemistry discipline, even if unconsciously, or not systematized.

6   Conclusions

The analysis of disciplines confirms the perception of the importance of tutors in conducting the teaching-learning process in distance education.

Feedback is not a different interactive process if compared to classroom education, requiring only to observe the differences about the temporality of message exchanging. Most tutors adopt the sandwich feedback model, however the feedback ladder model seems to be more suitable for favoring significant learning including in their steps actions that contribute to this process (clarify, value, question, suggest).

Feedbacks promote a continuous cycle of building collective knowledge, when every interaction knowledge and experiences are shared promoting a new meaningfulness to prior knowledge, leading to inter-cognition. According to the good appointment from Bicalho and Oliveira [3], this sequence of feedback adds other elements, other interpretative repertoires, strengths the collective argument and shapes the other contributions, so that the knowledge and experience cam bel externalized and internalized in each individual, therefore contributing to the meaningful learning process.

The discussion forum as building of knowledge in distance education is an excellent space of collective cognition (inter-cognition), through which, collaboratively, participants (teachers, tutors and students) can build the space changes from reflection and social construction of knowledge.

References

1. Perrier, G.R.F.: A importância do feedback nas atividades assíncronas do AVEA Moodle: discutindo o papel do tutor enquanto agente facilitador da aprendizagem significativa na EaD. 2013. 80 f. Trabalho de Conclusão de Curso (Especialização). Universidade Federal de Santa Catarina, Florianópolis (2013)
2. de Bicalho, R.N.M.: O processo de construção intersubjetivo do conhecimento em educação a distância. Dissertação (Mestrado). Universidade de Brasília, Brasília (2010)
3. de Bicalho, R.N..; de Oliveira, M.C.S.L.: O processo dialógico de construção do conhecimento em fóruns de discussão. Interface, Botucatu **16**(41), Junho 2012. http://www.scielo.br/pdf/icse/v16n41/aop2712.pdf
4. Ausubel, D.P.: Educational Psychology: A Cognitive View. Rinehart & Wilson, New York (1968)
5. Williams, R.L.: Preciso saber se estou indo bem: uma história sobre a importância de dar e receber feedback. Sextante, Rio de Janeiro (2005)
6. Lima, D.M.A., Alves, M.N.: O feedback e sua importância no processo de tutoria a distância. Revista Pro-Posições **22**(2), Campinas, São Paulo (2011)
7. Moreira, M.A.: Aprendizagem significativa. UNB, Brasília (1999)

Part VI
Validation, Testing, Prototyping and Simulation for Assessing TEL System Quality

Accuracy of Second Language Skills Self-assessment Through a Web-Based Survey Tool

Vincenzo Baraniello, Chiara Degano, Luigi Laura, María Lozano Zahonero, Maurizio Naldi and Sandra Petroni

Abstract A web-based tool is employed to conduct a survey and analyse the accuracy of self-assessment when L1 and L2 are genetically and typologically related: a Spanish class attended by Italian students. The survey includes a questionnaire and a test. A correlation analysis based on Goodman-Kruskal's Gamma index shows that the self-assessment of general reading skills is weakly correlated with the self-assessment of comprehension of a specialised text. The latter is even more weakly correlated with actual test results. Self-assessment in a context of cognate languages is a poor predictor of students' actual performances.

Keywords Second language · Self-assessment · Translation

V. Baraniello · C. Degano · M.L. Zahonero · S. Petroni
Department of Philosophy Literature Art, University of Rome Tor Vergata,
Via Columbia 1, 00133 Rome, Italy
e-mail: {baraniello,chiara.degano,sandra.petroni}@uniroma2.it,
lozano.zahonero@lettere.uniroma2.it

L. Laura
Department of Computer, Control, and Management Engineering "Antonio Ruberti",
University of Rome La Sapienza, Via Ariosto 25, 00185 Rome, Italy
e-mail: laura@dis.uniroma1.it

L. Laura
Research Center for Distance Education and Technology Enhanced Learning (DETEL),
Unitelma University, Rome, Italy

M. Naldi(✉)
Department of Civil Engineering and Computer Science, University of Rome Tor Vergata,
Via del Politecnico 1, 00133 Rome, Italy
e-mail: naldi@disp.uniroma2.it

© Springer International Publishing Switzerland 2016
M. Caporuscio et al. (eds.), *mis4TEL*,
Advances in Intelligent Systems and Computing 478,
DOI: 10.1007/978-3-319-40165-2_17

163

1   Introduction

Translation can be considered as a fifth skill, whose use is suggested in a computer-assisted language learning strategy [7]. This is the aim of the recently launched WALLeT project, which plans to create a collaborative platform where a computer-aided translation (CAT) tool is used in a second language learning environment [1, 2]. The project exploits the popular MediaWiki and its Translation extension translatewiki, which allows for embedded CAT functionalities and, in particular, to access the Translate Toolkit server, a Web server for translation memories.

In the framework of the WALLeT project, also the Google Forms tool has been employed to create and rapidly disseminate questionnaires and tests, which students can fill in both in the classroom and at home to test their capabilities and provide their feedback on learning methods and materials. The use of Google Forms in a learning context has already been reported, e.g., in [5].

A procedure typically employed in the classroom is a preliminary self-assessment of second language skills by the students themselves. The function of self-assessment and the critical nature of its accuracy are a long-standing issue (see the two papers [3] and [4], over ten years apart), but still deserve further investigation.

Since the late 1950's many scholars have highlighted that dissimilarities between L1 and L2 may correlate with L2 learnability through L1 interference or negative transfer [6, 8, 9]. Instead, it was long supposed that L1-L2 similarities had only a positive effect on learnability. However, it has been recently pointed out that formal similarities can cause several types of homograph interference when L1=Italian and L2=Spanish, two genetically and typologically related languages [10]. The accuracy of students' second language self-assessment capabilities (i.e., their over- or under-estimation) is then to be explored.

In this paper, we jointly investigate the two issues, namely self-assessment of second language skills and the influence of linguistic distance on self-assessment capabilities, through the use of the web-based Google Forms platform in a classroom where a CAT tool is employed for collaborative language learning. We show that students' self assessment of their general reading skills is weakly correlated with their self-assessment of specialised text comprehension, and in turn actual comprehension results are very weakly correlated with their self-assessment.

The paper is organized as follows. We describe the questionnaire employed in the classroom in Section 2, state the research questions and describe the methodology employed in Section 3, and report the results in Sections 4 and 5.

2   The Questionnaire

In the framework of the WALLeT project, students have been asked to fill in a questionnaire and take an associated test to assess their knowledge of their second language of choice. A module was devoted to assess their self-assessment capabilities. In this section, we describe that module.

The questionnaire was distributed to a Spanish class formed by 44 students enrolled in several undergraduate programmes; all of them filled in the questionnaire. The distribution of students by their domain of knowledge is shown in Fig. 1, where the labels in the legenda on the right side are ordered by decreasing frequency. As can be seen, though the largest group was that of students of Languages, a fairly wide variety of fields of study was represented.

The questionnaire section devoted to self-assessment included a text, approximately 100 words long, for each knowledge domain. Students were presented two texts, concerning respectively their own domain of knowledge (hereafter referred to as D1 for short) and a different domain of knowledge (hereafter we refer to the

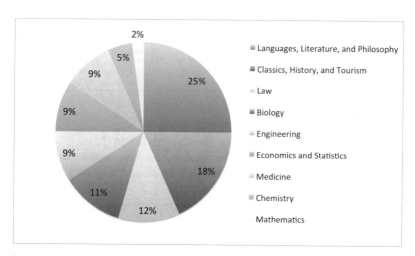

Fig. 1 Distribution of students by domain of knowledge

Table 1 Couples of domains

One's own domain of knowledge	Alien domain of knowledge
Mathematics	Geology
Physics	Law
Chemistry	Tourism
Geology	Mathematics
Biology	Engineering
Medicine	Languages
Agriculture and Veterinary Medicine	Economics and Statistics
Engineering	Biology
Classics, History, and Tourism	Chemistry
Languages, Literature, and Philosophy	Medicine
Law	Physics
Economics and Statistics	Agriculture and Veterinary Medicine

Table 2 List of questions

Q. ID	Question	Possible answers
	General	
1.1	What is your reading skill level?	A1 through C2 (as defined in the Common European Framework of Reference for Languages)
	Domain D1	
2.1	Did you understand the text?	0 (nothing) through 5 (all)
2.2	List the three specialised terms that most helped you understand the domain	Any three terms
2.3	Which text was the excerpt taken from?	Multiple choice question
2.4	Which is the intended audience?	Multiple choice question
2.5	Choose a sentence that belongs to the same text	Multiple choice question
2.6	Does the text contain any error?	Multiple choice question
	Domain D2	
3.1	Did you understand the text?	0 (nothing) through 5 (all)
3.2	Which domain of knowledge does the text belong to?	Multiple choice question
3.3	List the three specialised terms that most helped you understand the domain	Any three terms
3.4	Which text was the excerpt taken from?	Multiple choice question
3.5	Which is the intended audience?	Multiple choice question
3.6	Choose a sentence that belongs to the same text	Multiple choice question
3.7	Does the text contain any error?	Multiple choice question

second domain as an alien one, D2 for short), as shown in Table 1. After reading each text, they were asked to answer the questions reported in Table 2.

An overall score was computed for each student in both domains. For domain D1, one point was assigned for each correct answer to Q. 2.3 through 2.6 and 1/3 points for each correct term in Q. 2.2. The maximum score for domain D1 was 5. Instead, for domain D2, one point was assigned for each correct answer to Q. 3.2 and 3.4 through 3.7 and 1/3 points for each correct term in Q. 3.3, so that the maximum score was 6.

3   Research Questions

The questionnaire is instrumental in addressing some research issues concerning students' self-assessment capabilities. In this section, we state the research questions and describe the statistical approach taken to carry out the analysis.

As pointed out in the Introduction, the rationale for our analysis is that native speakers of a language may not assess correctly their comprehension capability for a second cognate language. In our case, we have examined the self-assessment capabilities of Italian students when reading a Spanish specialized text, using the responses to the questionnaire described in Section 2.

The list of questions shown in Table 2 basically contains three items:

 – a general self-assessment of the student's second language skills (Q. 1.1)
 – a self-assessment of the student's comprehension of the specialised text, for each of the two domains (Q. 2.1 and 3.1)
 – a comprehension test (Q. 2.2 through 2.6 + Q. 3.2 through 3.7)

The research questions we intend to address are the following

RQ1 Does the student's general self-assessment agree with his/her self-assessment for his/her own specialised domain of knowledge?

RQ2 Does the student's general self-assessment agree with his/her self-assessment for an alien specialised domain of knowledge?

RQ3 Is the student's self-assessment for his/her own specialised domain of knowledge correct?

RQ4 Is the student's self-assessment for an alien specialised domain of knowledge correct?

We look for answers to these research questions by correlating the answers to the questionnaire. Those answers are represented by either ordinal variables (Q. 1.1, 2.1, and 2.3) or discrete numerical variables (the numerical scores obtained from Q. 2.2 through 2.6, and Q. 3.2 through 3.7), which are ordinal variables as well, since the score is obtained through a conventional assignment.

In order to measure the correlation between those answers, we employ the Goodman-Kruskal's Gamma index, which is the most suitable to measure correlation between two ordinal variables (see Equation 32.1 of [11])

$$G = \frac{n_c + n_d}{n_c - n_d},$$ (1)

where n_c and n_d are respectively the number of concordant pairs (i.e., the number of pairs of cases ranked in the same order on both variables) and number of reversed pairs (i.e., the number of pairs of cases ranked in reversed order on both variables), as computed on the contingency table. Values range from -1 (which represents a perfect negative correlation) to +1 (which represents a perfect agreement), while a value of zero indicates the absence of correlation.

We also use the Gamma index to perform a significance test about the presence of correlation. Namely we contrast the null hypothesis that $G = 0$ (i.e., there is no correlation between the variables) with the alternative hypothesis that $G > 0$ (i.e., there is a positive correlation). In order to carry out the test we compute the test statistic z (see Equation 32.2 of [11])

$$z = G\sqrt{\frac{n_c + n_d}{n(1 - G^2)}},$$ (2)

where n is the sample size. The null hypothesis can be rejected if $z > z_q$, z_q being a chosen quantile of the standard normal distribution (typically $q = 0.05$ or $q = 0.01$, so that respectively $z_{0.05} = 1.65$ and $z_{0.01} = 2.33$).

4 Agreement of Self-assessments

After stating the research questions and the methodology of analysis in Section 3, in this section we examine the research questions RQ1 and RQ2 to see if the student is consistent in his/her self-assessment at a general level and when s/he has to confront a real text.

Since the test is made of a written text to be read, for the research question RQ1 we compare the answer to Question 1.1 with the answer to Q. 2.1. Instead, for the research question RQ2 we compare the answer to Q. 1.1 with the answer to Q. 3.1. For both cases we compute the Gamma index G and the test statistic z, as described in Section 3.

We first show the contingency tables for the two cases (Domains D1 and D2) in Table 3. The numbers highlighted in bold are the highest correlation found for each declared reading skill level. We would expect the array of bold figures to approximately form a diagonal across the table, since students with a higher level should be more confident of understanding the proposed text. We see immediately that some disagreement exists between the two self-assessment answers, since the diagonal scheme is not followed in either domain.

The results of the application of the Goodman-Kruskal test are shown in Table 4. As can be seen, the Gamma index is approximately equal for the two domains, sitting roughly halfway between the no-correlation value ($G = 0$) and the perfect agreement ($G = 1$) and exhibiting therefore a weak correlation.

The figures obtained for the hypothesis test tell us that the null hypothesis (that there is no correlation between the self-assessment provided for the general level of second language knowledge and that provided for the specialised text at hand) can be rejected at the 5% significance level but not at the tighter 1% level.

5 Self-assessment vs Actual Comprehension

In this section we examine the research question RQ3 and RQ4, i.e., the agreement between the specific self-assessment submitted by the student for the specialised texts provided and his/her actual score.

For the research question RQ3 we compare the answer to Q. 2.1 with the score obtained from Q. 2.2 through 2.6. Instead, for the research question RQ2 we compare the answer to Q. 3.1 with the score obtained from Q. 3.2 through 3.7. For both cases we compute the Gamma index G and the test statistic z, as described in Section 3.

We first show the contingency tables for the two cases (Domains D1 and D2) in Table 5 and Table 6 respectively. The numbers highlighted in bold are the highest correlation found for each declared comprehension level (i.e., the highest value in the row). We see that the expected diagonal pattern is slightly visible in the domain D1, but not at all in D2.

Table 3 Contingency tables for self-assessments agreement

	Domain D1						Domain D2					
	0	1	2	3	4	5	0	1	2	3	4	5
A1	0	0	1	**2**	0	1	0	0	**3**	1	0	0
A2	0	1	**6**	**6**	2	0	0	**5**	**6**	3	1	0
B1	0	0	4	**6**	**7**	2	0	4	3	**6**	5	1
B2	0	0	0	0	**4**	1	0	0	0	**2**	**3**	0
C1	0	0	0	0	**1**	0	0	0	0	0	0	**1**
C2	0	0	0	0	0	0	0	0	0	0	0	0

Table 4 Goodman-Kruskal test results for the correlation of self-assessments

Domain	n_c	n_d	G	z
D1	377	111	0.545	2.17
D2	398	120	0.537	2.18

Table 5 Contingency table for self-assessment-to-score correlation in domain D1

Self-assessment\Score	0	$0.\overline{3}$	$0.\overline{6}$	1	$1.\overline{3}$	$0.\overline{6}$	2	$2.\overline{3}$	$2.\overline{6}$	3	$3.\overline{3}$	$3.\overline{6}$	4	$4.\overline{3}$	$4.\overline{6}$	5
1	0	0	0	**1**	0	0	0	0	0	0	0	0	0	0	0	0
2	0	0	0	1	1	0	**5**	0	0	1	1	0	1	0	0	1
3	0	0	0	2	0	0	0	**4**	1	3	0	2	1	0	0	1
4	0	1	0	1	0	0	1	0	**4**	2	0	0	1	0	2	2
5	0	0	0	0	0	0	0	0	0	**2**	1	0	0	0	0	1

Table 6 Contingency table for self-assessment-to-score correlation in domain D2

Self-assessment\Score	0	$0.\overline{3}$	$0.\overline{6}$	1	$1.\overline{3}$	$0.\overline{6}$	2	$2.\overline{3}$	$2.\overline{6}$	3	$3.\overline{3}$	$3.\overline{6}$	4	$4.\overline{3}$	$4.\overline{6}$	5	$5.\overline{3}$	$5.\overline{6}$	6
1	0	0	0	0	0	1	1	1	1	**3**	1	0	0	0	1	0	0	0	0
2	1	0	0	1	2	0	0	0	1	1	0	1	1	1	1	1	0	0	0
3	0	0	0	0	0	0	1	0	**2**	**2**	0	1	**2**	**2**	0	1	0	1	0
4	0	0	0	0	0	0	2	1	0	0	2	0	**3**	1	1	1	0	0	0
5	0	0	0	0	0	0	0	0	0	0	0	0	0	1	0	?1	0	0	0

Table 7 Goodman-Kruskal test results for the correlation between self-assessment and score

Domain	n_c	n_d	G	z
D1	432	217	0.331	1.35
D2	467	256	0.292	1.26

The results of the application of the Goodman-Kruskal test are shown in Table 7. As can be seen, the Gamma index is approximately equal for the two domains (again slightly lower for D2), but is quite closer to 0 than to 1, indicating therefore a very weak positive correlation. The test statistics z is lower than both the quantile thresholds, so that the null hypothesis that no correlation exists cannot be rejected even at the 5% significance level.

6 Conclusions

A questionnaire has been prepared and distributed through Google Forms to a class of Italian students learning Spanish, with the aim of checking their self-assessment capabilities.

The results, analysed through the Goodman-Kruskal Gamma index, show that the levels of reading comprehension declared in general and for a specialised text are weakly correlated: when dealing with a specialised text, students are led to deviate even significantly from their general self-assessment.

In addition, their actual reading skills exhibit an even weaker correlation with their declared self-assessment: students do not appear to be aware of their actual reading performance. Italian students of Spanish think they know Spanish better than they actually do.

These early results provide support to carry out a future comparative analysis between cognate and non-cognate languages, as well as between general and specialized languages, to test hypotheses concerning the linguistic distance effect.

References

1. Baraniello, V., Degano, C., Laura, L., Lozano Zahonero, M., Naldi, M., Petroni, S.: The WAL-LeT project (Wiki Assisted Language Learning and Translation): bridging the gap between university language teaching and professional communities of discourse. In: The Eighth International Conference in Discourse, Communication and the Enterprise (DICOEN 2015), Naples, June 11–13, 2015
2. Baraniello, V., Degano, C., Laura, L., Lozano Zahonero, M., Naldi, M., Petroni, S.: A wiki-based approach to computer-assisted translation for collaborative language learning. In: Li, Y., Chang, M., Kravcik, M., Popescu, E., Huang, R., Kinshuk, Chen, N.-S. (eds.) State-of-the-Art and Future Directions of Smart Learning. Lecture Notes in Educational Technology, pp. 369–379. Springer, Singapore (2016)
3. Blanche, P., Merino, B.J.: Self-assessment of foreign-language skills: Implications for teachers and researchers. Language Learning **39**(3), 313–338 (1989)
4. Brantmeier, C., Vanderplank, R., Strube, M.: What about me?: Individual self-assessment by skill and level of language instruction. System **40**(1), 144–160 (2012)
5. de la Fuente Valentín, L., Pardo, A., Delgado Kloos, C.: Using third party services to adapt learning material: A case study with google forms. In: Cress, U., Dimitrova, V., Specht, M. (eds.) Learning in the Synergy of Multiple Disciplines. Lecture Notes in Computer Science, vol. 5794, pp. 744–750. Springer, Heidelberg (2009)
6. Dechert, H.W., Brüggemeier, M., Fütterer, D.: Transfer and Interference in Language: A Selected Bibliography, vol. 14. John Benjamins Publishing (1984)
7. Garcia, I., Pena, M.I.: Machine translation-assisted language learning: writing for beginners. Computer Assisted Language Learning **24**(5), 471–487 (2011)

8. Gass, S.M., Selinker, L.: Language Transfer in Language Learning: Revised Edition, vol. 5. John Benjamins Publishing (1992)
9. Lado, R.: Linguistics Across Cultures: Applied Linguistics for Language Teachers. University of Michigan Press (1957)
10. Lozano Zahonero, M.: Los verbos de apoyo. In: San Vicente, F. (ed.) GREIT: Gramática de Referencia de Español Para Italófonos, vol. II, pp. 801–834. CLUEB (2013)
11. Sheskin, D.J.: Handbook of Parametric and Nonparametric Statistical Procedures. CRC Press (2003)

LTI for Interoperating e-Assessment Tools with LMS

Antonio J. Sierra, Álvaro Martín-Rodríguez, Teresa Ariza,
Javier Muñoz-Calle and Francisco J. Fernández-Jiménez

Abstract Learning Management Systems (LMS) are widely used nowadays because they let the users organize the learning process through courses, composed by documents, homework and exams, and it allows the teacher to follow students' progress. However LMS can't evaluate a student project due to its complexity. It can be solved by using external tools prepared for this specific task, losing the capability to track the student progress. Therefore the need of integration between them appears. To achieve a code project's assessment, we have developed an external tool interoperating with LMS using Learning Tools Interoperability (LTI).

Keywords Learning Management Systems · LTI · e-assessment · Cooperative/collaborative learning · System · Distributed learning environments

1 Introduction

Learning Management Systems (LMS) are a common software application in high education institutions, in order to have a regular communication between instructors and learners, sharing easily content and allowing learning process tracking. Learning process can be improved thanks to different capabilities that LMS can offer, like being able to share documents, create groups, assign roles, accomplish tasks or activities, or even make test exams. However there are more complex tasks that can't be achieved by these LMS. These tasks can be solved by creating specific external applications instead.

Our scenario is the subject named Fundamentals of Programming II (FP2), in Degree in Telecommunications Engineering Technology (DTET), that is taught at the

A.J. Sierra(✉) · Á. Martín-Rodríguez · T. Ariza · J. Muñoz-Calle · F.J. Fernández-Jiménez
Universidad de Sevilla, Sevilla, Spain
e-mail: {antonio,matere,javi,fjfj}@trajano.us.es, btc@hotmail.es

© Springer International Publishing Switzerland 2016
M. Caporuscio et al. (eds.), *mis4TEL*,
Advances in Intelligent Systems and Computing 478,
DOI: 10.1007/978-3-319-40165-2_18

173

Higher Technical School of Engineering (ETSI) at the University of Seville (US). Students have to code two projects, one using C language, and other using Java.

We want to evaluate student-code projects, but LMS can't execute code nor evaluate it. Even if it could, it wouldn't allow it, because malicious code would have a way to run into the LMS host. The solution with execution code into LMS would become a serious security problem. We could use an external tool to evaluate student-code projects, but a standalone external application wouldn't benefit from the user data contained in the LMS, and it wouldn't be able to send in any information after the user finishes using the application. Consequently, our best solution is to integrate the application, usually named tool, into the LMS.

In this scenario, we've to create a tool which lets us to evaluate our student's code, by compiling it and then executing, in order to compare the results to the expected ones. Because the tools to evaluate these student projects already exist, we created a tool which allows us to accomplish our goal using them. In order to integrate our tool into the LMS, we decided to use Learning Tools Interoperability (LTI) (IMS Global Learning Consortium, 2010) because between the different existing ways to achieve it, LTI is enough to make things work correctly, and is widely supported by learning environment. Just as example, Microsoft made Office products LTI 1.0 compliant (Microsoft, 2015).

Specifically, we have to evaluate student code using a new tool, but LMS don't allow running code, and it would be a security problem towards the system. This could be solved by executing the code within an external server, so integration between the tools running into this server and the LMS would be necessary. Starting from the different ways to complete this tool integration, we show how LTI works and which does it offer us, and we have developed a tool that uses Basic LTI. Finally, we show different test to which the tool has been undergoing, resulting in a working external tool integrated into a LMS.

2   Interoperability Alternatives

Integration of external tools can be achieved in different ways, each one with different features because of the architecture of the approach. The most known of these approaches are: LTI, Apache Wookie, and Group Learning Uniform Environment (GLUE!). Besides, LTI can be split in two different versions, being the first one Basic LTI, and the second one, just LTI. After them, there're other approaches less known, like: GSI Architecture, PoEML-based architecture, Sloodle, Gridcole, MoodleRooms or CCSI.

Apache Wookie (Apache Software Foundation) is an approach able to integrate W3C Widgets, and combined with Apache Shinding, it can work with OpenSocial compliant tools. To work, it needs a Wookie server, which is a standalone application which provides REST-based contract to access our deployed widgets within it. Besides, to integrate our widgets into the learning platform, a plug-in that would show available widgets would need to be developed for this platform. This plug-in would be in charge of setting-up and configure widgets instances also. It's not widely used, but W3C Widgets are standard. Considering that it can be combined with OpenSocial it presents a high growth potential, because it could integrate all existing applications that already are compliant with these standards. As an improvement, Wookie can work inside another server applications like Tomcat, expanding possibilities by not demanding an special server. However, this option needs also an MySQL database.

Next, GLUE! (Intelligent & Cooperative Systems Group (GSIC)) aims to take advantage of existing interfaces of platforms and tools. Made to support a basic integration of them, the development of an adaptor for each new tool or platform that is aimed to connect using GLUE! (Alario-Hoyos & Wilson, 2010). GLUElet Manager is the base component of the architecture. It stores a list about the available tools, and information about them, as well as manages the process of setting-up instances. Besides it, on the one hand there are Implementation Model Adaptors, which deal with the tools by creating tools instances thanks to their knowledge of tool contracts. On the other hand, VLE Adaptors are based on the platform extension interface. They can be installed as any other extension into the platform, and provide an interface for the users in order to interact with the integrated tools.

3 Learning Tools Interoperability (LTI)

Born in 2010, LTI (IMS Global Learning Consortium, 2010) is a standard which allow integrating tools into another platforms. In its very first version, the protocol only allowed launching the external tool, but these capabilities were improved with each new version of the standard until now.

The scenario in LTI is composed by the following actors, a student, a teacher, a LMS (as the common learning platform for the teacher and the student), and a tool (which is desired to be accessed and integrated from our LMS by the teacher).

But, what does it mean to be integrated? When both the human actors are logged into their common LMS, their actions over it are identified and processed under some parameters: a role (teacher / student), a name, an identifier... All of them allow it to customize the learning and use experience as they perform actions over the platform. Integration of new tools using Basic LTI (a.k.a. LTI 1.0) aims to let them to take advantage of that kind of data, in order to customize experience as well. To achieve it, the standard follows the steps shown in the Figure 1:

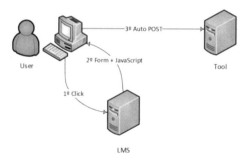

Fig. 1 Basic LTI Launch Steps

First of all, the teacher or administrator of the LMS, has to configure the link that will take the user to the tool. Supposing it to be already done, and the user connected into the LMS:

1. The user clicks over the LTI link. At this moment, the LMS:
   (a) Creates a new message, which contains:
      (i) User and context data, that was configured in the link setup to be shared, filtered by privacy policies from the user.
      (ii) A pre-shared key, which let the tool know who is accessing it.
   (b) Signs the message, using OAuth 1.0 (OAuth) framework, based on a pre-shared secret.
   (c) Adds the generated sign to the message.
   (d) Adds a JavaScript code, which will make the user to auto-POST a new request when he receives the message.
2. Then, the LMS sends the reply to the user.
3. The user, automatically POSTs is to the external tool thanks to the JavaScript code.
4. The tool process the request:
   (a) It takes the key from the message, and look for the associated secret in their database or configuration data.
   (b) Signs the message (avoiding the received sign) using this secret, and checks if the generated sign matches with the received one. If so, the received request is valid, if not, the launch can't be performed.

After being launched, the tool behaves like a normal web app, depending of its code and functions, but it has the additional data about the user that was received in the LTI request. In further versions of LTI, capabilities are expanded: the tool can return an integer to the LMS as result of its execution; REST services can be offered between both ends.

4   Related Work and Context

In 2010-2011 the first course in Degree in Telecommunication Technology Engineering (DTTE) at the University of Seville were established within the Department of Telematics Engineering with Fundamental of Programming I and Fundamental of Programming II. Due to the foundation of European Higher Education Area (EHEA) is based on student's work and labs, the establishment in these subjects have been applied by Project Based Learning (PBL) techniques (Sierra, 2012) (Sierra, 2013) (Sierra, 2014).

The purpose of these subjects is to establish the principles of computer programming. Every student has to complete two course projects. By one side, one project consists of an application in C programming language, comprising the following steps: understanding the problem, designing of the program, and subsequent testing. By other side, the other project consists of an application in object oriented paradigm, in Java programming language.

At present, evaluation support tools are used in nearly all education fields. We consider e-assessment as any assistance to an evaluation process by use of information technology and communication (ICT). This area covers tools for designing assessments to automatic correction, including monitoring students. Depending on the nature of the assistance we can define two types of activities: computer aided evaluation, where the evaluation is delegated in part to ICT or computer-based assessment when all is done by the evaluation tools. The computer-aided evaluation have great benefits, among which stand out: higher computing power, lower cost in the evaluation, greater accuracy is obtained, greater flexibility, immediate response, and more complex corrections.

There are other tools to automate all or part of the task of evaluating software projects in the field of education (Sierra, 2013) (Sierra, 2014). At the University of Seville the most popular method for evaluation assistance is Virtual Learning platform (Sierra, 2013). This platform offers a rich set of functionality such as shared use of files, ad serving, publication of ratings, use of calendars for planning assignments and publishing forums and so on. But, student's project code assessing is very complicated because it is necessary compiling and running a test suite. Currently there are about 300 students in the course, and so the individual correction for each of the work is discarded. So, it is necessary using another tool.

In previous works we have developed tools for validation software project to help in testing and developing student's project. The assessment has two aspects. On one hand, the project must comply with the specifications given to the student. Furthermore, students must pass a test in the computing facility. Students must changes the Project in a prescribed time.

First, we have developed a simple tool for testing (Sierra 2012). This tool compiles and runs student's code to check its proper implementation. The tool generates an output. It helps student to improve the project. The main disadvantage is this tool must be run locally. Scripts can be used for this purpose. However there are still several aspects to address. Student must modify his project and overcome some tests. After exam, it is necessary automatic assessing with the new requirements and generating scores automatically and getting results, all this with a friendly user interface.

To address some of these limitations, we applied client/server architecture and developed an e-assessment Java tool with a friendly user interface, for automatic C-project-code assessment. This tool enables students to compile and execute code, and check assessment tests remotely. The teacher can access to generated results stored in a data base. Students can verify the correct operation of its project by tool's client. Client connects to tool's server that processes requests and returns a result (Sierra 2014).

5   Developed Tool

We've developed a tool(Martín-Rodríguez, 2015) in order to be integrated into the LMS used at the University Of Seville. This platform is Blackboard Learn, LTI 1.0 compliant. Then, we need an LTI implementation to use at our tool. We aimed to run our tool into a Linux server, preferably using Java than PHP. With these requirements, the chosen option was BLTI-Sandwich, in order to make our tool LTI 1.0 compliant, because it's a Java implementation which can be easily used by building it and importing to our project.

The tool needs to present a simple interface, but with different options depending on which kind of user it is: learner or instructor. If the user is recognized as a learner, the interface must show the option to deliver the student project; if the user is an instructor, the interface must show the status of the tool (enabled/disabled, where it is going to store the learner files, etcetera) as well as being able to enable and disable the tool. Finally, we've added the capability to modify the delivery password at this interface. The password is a method used by the teachers in order to ensure that only the students that are present in the class can deliver the project, avoiding that students at home can use the tool without it. We can found the complete use cases in both next Fig. 2 and Fig. 3:

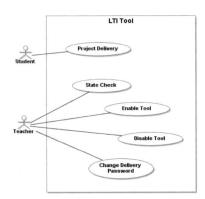

Fig. 2 Use cases   **Fig. 3** Use case "Deliver Project" Detailed

This is achieved having a servlet as entry point, being the first step of all to check if the request type is POST. If it is, the tool proceeds checking if it is a valid LTI request, to finally send the user, depending on his role, to his corresponding interface, that is made using JSP technology, as we can see in Fig. 4.

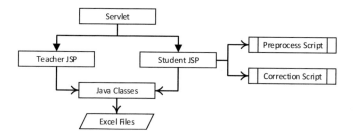

Fig. 4 Tool Composition Diagram

Both of these Java Server Pages (JSP) use some precompiled Java classes in order to simplify the error treatment and other task like the access to Excel files, where marks obtained from the correction task are stored. Besides, the student side employs two types of script:

- Preprocess script: it aims to prepare the student project to be evaluated. In our case, doing:

  — A backup of the file delivered by the student.
  — Unzip the file, so the real files contained inside it can be evaluated.
- Evaluation script: for each section which the student has to complete in the project, there's a script that will check if the result is the expected one, by making a test. This task is executed as following:

  — The script compiles the student code.
  — It executes the generated binary, giving it as parameter the evaluation files: configuration and input files. This task is done without security considerations, since student programming skills are basic at this point. However, this point is proposed at the end of this document as a future development line.
  — It compares the student generated output with the expected one. If they match, the section is evaluated as passed.

The mechanism described for the evaluation scripts also let us create complex evaluation task, if they can be checked by using a script. Specifically, one of the sections evaluated in one of the projects, is whether the student application has memory leaks that aren't released at the execution end, after allocating them. This is achieved by comparing a file, generated with an own created tool that is executed by the section script, with the amount of allocated bytes at the end of the execution. If the file contains a 0, the student application has released all the allocated memory and is correct.

Therefore, when a student uses the tool, the complete workflow could be described as in the following Fig. 5.

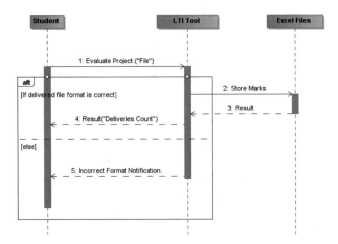

Fig. 5 Student Use Workflow Diagram

When all this have been executed, the tool stores the student marks into two different Excel files: one of them is just filled with all the students UVUS (US IDs) and their marks, which aims to be imported directly into the LMS to update the student scores; the other one, has the complete description about what did score the student in each section too, being this way more useful for the teacher. To achieve this, the application uses Apache POI library (Apache Software Foundation), which allow us to work with Microsoft Office files.

6   Conclusions

We have developed an external tool interoperating with LMS using LTI to achieve a code project's assessment. LMS are widely used nowadays because they let the users organize the learning process through courses, composed by documents, homework and exams, and it allows the teacher to follow students' progress. As future development lines, a further analysis and implementation of security measures at the tool host are recommended, in order to avoid data leaks from the host, as well as any type of damage to either the operative system or the tool.

References

1. Microsoft: OneNote Class Notebook and Office Mix announce new LMS integration features, October 26, 2015. https://blogs.office.com/2015/10/26/onenote-class-notebook-and-office-mix-announce-new-lms-integration-features/ (accessed October 28, 2015)
2. Intelligent & Cooperative Systems Group (GSIC): GLUE!. http://www.gsic.uva.es/glue/ (accessed September 10, 2015)

3. Alario-Hoyos, C., Wilson, S.: Comparison of the main alternatives to the integration of external tools in different platforms. In: Proceedings of the International Conference of Education, Research and Innovation, Madrid, Spain (2010)
4. IMS Global Learning Consortium: Basic LTI Implementation Guide Version 1.0 Final, Mayo 17,2010. http://www.imsglobal.org/lti/blti/bltiv1p0/ltiBLTIimgv1p0.html (accessed Septiembre 5,2015)
5. OAuth. http://oauth.net/ (accessed October 20, 2015)
6. Sierra, A., Ariza, T., Fernandez, F.: Establishment EHEA for telecommunication technologies engineering degree. In: Global Engineering Education Conference (EDUCON), Marrakesh (2012)
7. Sierra, A.J., Ariza, T., Fernández, F.J.: PBL in Programming Subjects at Engineering. Bulletin of the IEEE Technical Committee on Learning Technology **15**(2), 18–21 (2013)
8. Sierra, A., Ariza, T., Fernández-Jiménez, F.: E-assessment tool within PBL in programming subjects at engineering. In: EDULEARN 2014, Barcelona (2014)
9. Sierra, A., Ariza, T., Fernández-Jiménez, F., Madinabeitia, G., Muñoz-Calle, J., Martínez-Gotor, A.: PBL e-assessment client/server tool at programming subjects for engineering. In: ICERI 2014, Seville (2014)
10. Martín-Rodríguez, Á.: Comunicación Entre Herramientas Basadas en LTI y Sistemas de Gestión de Aprendizaje. http://bibing.us.es/proyectos/abreproy/90452/fichero/TFG+%C1lvaro+Mart%EDn+Rodr%EDguez.pdf (retrieved)
11. Sierra, A., Ariza, T., Fernandez, F., Madinabeitia, G.: Tool for Validation Software Projects in Programming Labs. International Journal of Engineering Pedagogy (iJEP) **2**(2) (2012)

Antecedents of Achievement Emotions: Mixed-Device Assessment with Italian and Australian University Students

Daniela Raccanello, Margherita Brondino, Monique Crane and Margherita Pasini

Abstract We focused on cross-cultural comparisons on achievement emotions, as those emotions associated to learning activities or outcomes. An advantageous way to investigate them is using on-line assessment tools, enabling users to gather data in economic ways. We involved 206 Italian and Australian first-year university students, who were administered an on-line questionnaire measuring challenge and threat appraisals, two emotion regulation strategies, and ten achievement emotions. The results indicated cross-cultural differences for threat, reappraisal, positive deactivating, negative activating, and negative deactivating emotions. Path analyses showed that challenge and threat appraisals, and reappraisal and suppression strategies, coherently predicted achievement emotions, with some exceptions, invariantly across groups. These findings confirm the usefulness of using on-line assessment, and inform on cross-cultural comparisons on achievement emotions, documenting differences in their mean levels, but supporting the universality of the relationships between antecedents and subsequent emotions.

Keywords Achievement emotions · Mixed-device assessment · Cross-cultural comparisons

1 Introduction

At present, there is a burgeoning increase in the diffusion of advanced learning technologies involving the most disparate aspects of people's life. In a parallel

D. Raccanello(✉) · M. Brondino · M. Pasini
Department of Philosophy, Education and Psychology,
University of Verona (Italy), Lungadige Porta Vittoria, 17, 37129 Verona, Italy
e-mail: {daniela.raccanello,margherita.brondino,margherita.pasini}@univr.it

M. Crane
Department of Psychology, Building C3A,
Macquarie University, Sydney, NSW 2109, Australia
e-mail: monique.crane@mq.edu.au

© Springer International Publishing Switzerland 2016
M. Caporuscio et al. (eds.), *mis4TEL*,
Advances in Intelligent Systems and Computing 478,
DOI: 10.1007/978-3-319-40165-2_19

183

way, some of these technologies have begun to be used to assess how people perceive, elaborate, and react to everyday events, also in relation to learning environments such as school and academic settings [9]. In this line, the use of on-line assessment tools to measure how people feel with respect to specific contexts has the advantage to combine the strength of traditional and innovative instruments. On the one hand, the use of classical self-report measures, whose reliability and validity has already been documented, through technological modalities make it simpler for example to compare face-to-face and technological learning environments [22]. On the other hand, presenting assessment tools via on-line systems enables to go beyond constraints typical of some studies related for example to time and procedural limitations. Therefore, in this work we present a research study on the use of an on-line assessment tool proposed to evaluate some emotional dimensions linked to specific learning contexts. The questionnaire proposed presents some characteristics particularly valuable for Technology Enhanced Learning (TEL) environments [22], such as common-use language, short duration, and simple answer format, a modality linked to low anxiety [19].

We focused on achievement emotions experienced by first-year Italian and Australian university students with reference to a specific course they were attending. According to the control-value theory [16, 18, 19], achievement emotions are those emotions linked to learning activity or outcomes, differentiated for valence (positive, negative emotions) and activation (activating, deactivating emotions). Beyond assuming and documenting the nature of the proximal antecedents of achievement emotions, in terms of beliefs on people's control and task-value, the model describes how they influence, also through mediating and moderating factors, achievement. Positive activating emotions (i.e., enjoyment, hope, pride) would be beneficial for learning, while negative deactivating emotions (i.e., boredom, hopelessness) would be detrimental; for positive deactivating (i.e., relief, relaxation) and negative activating (i.e., anxiety, anger, shame) emotions, however, some contradictory results have been found.

Investigating the nature of antecedents of achievement emotions is a useful way to understand which factors can be responsible of both students' emotions and achievement. We focused on two types of antecedents partially disregarded, cognitive appraisals and emotion regulation strategies. On the one hand, challenge appraisals refer to 'a focus on the potential for gain or growth inherent in an encounter' (p. 33), while threat appraisals to 'harms or losses that have not yet taken place but are anticipated' (p. 32) [12]. Given their differentiation in terms of positive versus negative expectations for an event, some data have documented how they predict positive versus negative emotions, for example considering excitement or anxiety in sport contexts [3, 11]. On the other hand, emotion regulation, 'as the ability to decrease, maintain or increase one's emotional arousal to facilitate engagement with the context' [14, p. 624], is strictly connected to achievement [5], but less is known on its role as predictor of achievement emotions. Among different emotion regulation strategies, cognitive reappraisal ('a form of cognitive change that involves construing a potentially emotion-eliciting situation in a way that changes its emotional impact') is one of the most effective ones, also in terms of enhanced achievement, while expressive suppression ('a form of response modulation that involves inhibiting ongoing emotion-expressive behavior') has been frequently labelled as ineffective [1, 10, p. 349].

Our general aim was to explore differences related to achievement emotions in Italian compared to Australian first-year university students using on-line instruments, taking into account that links between achievement emotions and culture has been only partially explored [8, 13, 17] and that cross-cultural comparisons have been suggested as one of the directions for future research on achievement emotions [6, 19]. For both groups, the emotions referred to similar Psychology courses they were attending, focusing on the studying setting, in order to take into account the context-specificity of achievement emotions documented by the control-value theory [16, 18].

First, we examined the reliability of the instruments, considering that the questionnaire used to assess achievement emotions was adapted for the first time in English. Second, we explored differences in the mean values of ten different achievement emotions, considering their type (positive activating, positive deactivating, negative activating, and negative deactivating emotions). We also checked for differences in the mean values of other two constructs related to emotions, i.e., challenge and threat appraisals and emotion regulation strategies. Third, we tested a model of relationships between challenge and threat appraisals and emotion regulation strategies (reappraisal and suppression strategies), on the one hand, and achievement emotions, on the other hand, considering the first two factors as antecedents. Coherently with the literature [1, 3, 10, 11], we expected positive emotions to be predicted positively by challenge and reappraisal strategies and negatively by threat and suppression strategies, and vice versa for negative emotions, separately for each group. Finally, we explored the invariance of the models across the two groups (Italian, Australian students), expecting the models to be invariant, on the basis of the literature [18, 19].

2   Method

2.1   Participants

We involved two samples of students from Italy and Australia (n = 206, 74% females). The Italian sample included 91 first-year students attending the course of General Psychology at the University of Verona (mean age = 22.03 years, SD = 5.28, range: 19-50 years). The Australian sample included 115 first-year students attending the course Introduction to Psychology at Macquarie University (mean age = 20.14 years, SD = 6.12, range: 17-55 years). All the students signed a consent form.

2.2   Material and Procedure

The study battery was administered remotely using Qualtrics software with a common Internet browser (Windows, Linux, Mac) and any device (computer, tablet, mobile, etc.). Qualtrics software enables users to develop online surveys and run different types of online data collection. The use of a mode switching in a mixed-device survey design, i.e. the possibility to choose any device to answer, could be particularly useful in this kind of surveys to maximize participation and minimize attrition [24].

Participants were emailed a link to the experimental questionnaires. They viewed an information page with the description of the research, and were given the opportunity to provide consent via checking a box or to revoke consent. All the students participated on a voluntary basis and were assured about anonymity. Then they completed a demographics questionnaire and the psychological measures, referred to challenge and threat appraisals, emotion regulation strategies, and achievement emotions related to studying for the course General Psychology or Introduction to Psychology.

2.2.1   Challenge and Threat Appraisals

We assessed appraisals with eight items [7], four for challenge (e.g., *Studying Introduction to Psychology/General Psychology seems like a threat to me*) and four for suppression (e.g., *Studying Introduction to Psychology/General Psychology might reveal my weaknesses*), to be evaluated on a 7-point Likert scale (1= *strongly agree* and 7 = *strongly disagree*).

2.2.2   Emotion Regulation Strategies

We measured two strategies of emotion regulation, reappraisal and suppression, adapting two subscales from the Emotion Regulation Questionnaire (ERQ) [10]. We proposed six items, three for reappraisal (e.g., *I control my emotions by changing the way I think about the situation I'm in*) and three for suppression (e.g., *I control my emotions by not expressing them*), to be evaluated on a 7-point Likert scale (1= *strongly agree* and 7 = *strongly disagree*).

2.2.3   Achievement Emotions

We assessed ten achievement emotions with a self-report instrument developed on the basis of Pekrun's control-value theory [16, 17, 18, 19], the Achievement Emotions Adjective List, AEAL [2, 20, 21, 22, 23]. The questionnaire includes 30 adjectives related to three positive activating emotions (enjoyment, hope, pride), two positive deactivating emotions (relief, relaxation), three negative activating emotions (anxiety, shame, anger), and two negative deactivating emotions (boredom, hopelessness). The students were asked to indicate how they felt studying for a course they were attending, i.e. General Psychology for the Italian sample and Introduction to Psychology for the Australian sample, evaluating how much each word described their feelings on a 7-point Likert scale (1= *not at all* and 7 = *completely*). The order of the words was randomized and kept constant. The original version of the instrument is in Italian, and it was adapted in English using back-translation procedures.

2.3 Analysis Procedure

We calculated descriptive statistics, intercorrelations, and t-tests using SPSS version 21.0 for Windows. Level of significance was set at $p < .05$. We ran path analyses and multi-group analysis using Mplus version 6.11 [15]. There were missing data only for the Australian sample. To check if they were 'missing completely at random' (MCAR) we conducted the Little's MCAR test, $X2(468) = 495.19$, $p = .186$, which indicated that no identifiable pattern existed for the missing data. Missing values on predictor variables were calculated in Mplus using full information maximum likelihood (FIML) estimation. To explore the invariance of the paths across groups (Italian, Australian students), four models were conducted with multi-group analysis.

3 Results and Discussion

3.1 Reliability, Descriptive Statistics, Intercorrelations, and t-tests

We checked for reliability calculating α-values for challenge and threat appraisals, emotion regulation, and emotions. All α-values were higher than .64 (Table 1), indicating homogeneity for each construct. Therefore, students' responses on items concerning challenge and threat appraisal, emotion regulation strategies, and achievement emotion types were averaged together, separately by group (Italian, Australian).

Then, we conducted a series of independent sample t-tests, with group as the independent variable, and challenge and threat appraisals, emotion regulation strategies, and four achievement emotion types as dependent variables. We found significant differences for threat, reappraisal, positive deactivating, negative activating, and negative deactivating emotions, with higher scores for Australian compared to Italian students (Table 2).

Table 1 Intercorrelations and reliability for cognitive appraisals, emotion regulation strategies, and achievement emotion types for Italian and Australian students

	1	2	3	4	5	6	7	8
1. Challenge	.64	.15	.12	.06	.30**	.09	.15	-.01
2. Threat	-.05	.70	.03	.06	-.43***	-.40***	.63***	.55***
3. Reappraisal	.17	.04	.72	.15	.37***	.28**	-.14	-.12
4. Suppression	-.04	.08	.03	.86	-.11	-.12	-.03	.02
5. Positive activating emotions	.32**	-.34**	.15	-.01	.92	.76***	-.32***	-.38 ***
6. Positive deactivating emotions	.26*	-.45***	.26*	.06	.79***	.85	-.41***	-.38***
7. Negative activating emotions	.02	.44***	-.19	.10	-.25*	-.40***	.92	.76***
8. Negative deactivating emotions	-.10	.47***	-.18	.03	-.31**	-.35**	.65***	.88

Note. Correlations for the Italian sample are reported below the diagonal, while correlations for the Australian sample are reported above the diagonal. Alpha-values are reported along the diagonal. $*p < .05, **p < .01, ***p < .001$.

Table 2 Means (*M*), standard deviations (*SD*), and *t*-tests (degrees of freedom, *df*) for cognitive appraisals, emotion regulation strategies, and achievement emotion types for Italian and Australian students

	Italian sample		Australian sample		T-tests (df)
	M	SD	M	SD	
1. Challenge	5.72	.62	5.53	.87	1.81 (204)
2. Threat	2.74	.87	3.36	1.17	-4.18*** (204)
3. Reappraisal	4.74	.96	5.22	1.11	-3.29** (204)
4. Suppression	3.67	1.40	4.05	1.65	-1.74 (204)
5. Positive activating emotions	4.44	1.15	4.60	1.04	-1.01 (204)
6. Positive deactivating emotions	3.57	1.14	4.00	1.04	-2.85** (204)
7. Negative activating emotions	1.65	.59	3.51	1.03	-15.41*** (204)
8. Negative deactivating emotions	1.55	.65	3.75	.99	-18.31*** (204)

$*p < .05, **p < .01, ***p < .001$.

3.2 Path Analysis

Challenge positively predicted positive activating emotions for both samples, and positive deactivating emotions for the Italian sample (see Fig. 1 for the four models). Threat negatively predicted positive activating and deactivating emotions, and it negatively predicted negative activating and deactivating emotions, for both samples. Reappraisal positively predicted positive activating emotions for the Australian sample, and positive deactivating emotions for both samples; it negatively predicted negative activating emotions for both samples, and negative deactivating emotions for the Italian sample. Suppression negatively predicted negative activating emotions for the Australian sample. It is worth noting that the explained variance was higher for the Australian students compared to the Italian students in all cases except than for positive deactivating emotions. Finally, for each emotion type threat was the stronger predictor for both samples.

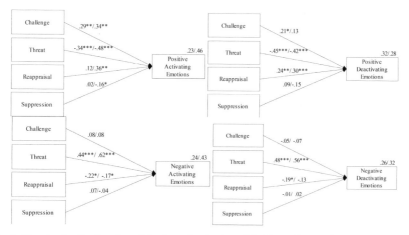

Fig. 1 The standardized paths of the hypothesized models for each emotion type, separately by group (Italian/Australian). Explained variances are reported next to the emotion variable. $*p < .05$, $**p < .01$, $***p < .001$.

In addition, we checked for invariance across the two groups (Italian, Australian), assessing simultaneously for the two groups the predicting role of cognitive appraisals and emotion regulation strategies on emotions. We found no statistically significant differences, supporting the invariance of the paths for Italian and Australian students.

4   Conclusions

Our findings document the reliability of using on-line assessment tools to compare emotional dimensions related to learning environments across cultures, through the use of a mixed-device approach which could have contributed to maximize data quality [24]. Filling gaps in the literature, we found both some appraisals and some emotions (mainly negative emotions) to be more intense for the Australian sample, but we also documented the invariance across groups in the paths of the models linking scarcely investigated antecedents such as cognitive appraisals and emotion regulation to achievement emotions. However, the generalizability of these patterns should be further explored in the future, to check for possible limitations related for example to sample size.

On the whole (with some exceptions), we found that challenge was related only to positive emotions, positively, while threat was related both to positive and negative emotions, respectively negatively and positively. Our findings are consistent with the recent work of Crane and Searle [4] that found challenge stressors (e.g., performing skilled tasks) were unique in their positive contribution to employee resilience three months later. In contrast, hindrance stressors (e.g., bureaucracy), that reflect threats and barriers to valued goals, reduced resilience and increased stress. Crane and Searle suggest that important in whether stressors generate adaptive outcomes is the meaning applied to those stressors. Stressors that are perceived as challenges become an opportunity for the conversion of stressors into opportunities for positive growth and resilience. In contrast, the perception of threat reduces such opportunities.

As regards emotion regulation strategies, reappraisal resulted related to all types of emotions, positively to positive emotions and negatively to negative emotions, while suppression only rarely. Beyond it theoretical relevance, acknowledging these trends has useful implications from an applied perspective, in terms of possible factors that can be enhanced to favour positive achievement emotions and discourage negative achievement emotions in specific learning contexts.

References

1. Augustine, A.A., Hemenover, S.H.: On the relative effectiveness of affect regulation strategies: A meta-analysis. Cognition and Emotion **23**(6), 1181–1220 (2009). doi:10.1080/02699930802396556
2. Brondino, M., Raccanello, D., Pasini, M.: Achievement goals as antecedents of achievement emotions: The 3 X 2 achievement goal model as a framework for learning environments design. Advances in Intelligent and Soft Computing **292**, 53–60 (2014). doi:10.1007/978-3-319-07698-0
3. Cerin, E.: Anxiety versus fundamental emotions as predictors of perceived functionality of pre-competitive emotional states, threat, and challenge in individual sports. Journal of Applied Sport Psychology **15**, 223–238 (2003). doi:10.1080/10413200305389
4. Crane, M.F., Searle, B.J.: Building resilience through exposure to stressors: The effects of challenges versus hindrances. Journal of Occupational Health Psychology (2016). http://dx.doi.org/10.1037/a0040064
5. Davis, E.L., Levine, L.J.: Emotion regulation strategies that promote learning: Reappraisal enhances children's memory for educational information. Child Development **84**, 361–374 (2013). doi:10.1111/j.1467-8624.2012.01836.x
6. DeCuir-Gunby, J.T., Williams-Johnson, M.R.: The influence of culture on emotions. In: Pekrun, R., Linnenbrick-Garcia, L. (eds.) International Handbook of Emotions in Education, pp. 539–557. Taylor and Francis, New York (2014)
7. Drach-Zahavy, A., Erez, M.: Challenge versus threat effects on the goal-performance relationship. Organizational Behavior and Human Decision Processes **88**, 667–682 (2002)
8. Frenzel, A.C., Thrash, T.M., Pekrun, R., Goetz, T.: Achievement emotions in Germany and China: A cross-cultural validation of the Academic Emotions Questionnaire-Mathematics (AEQ-M). Journal of Cross-Cultural Psychology **38**, 302–309 (2007). doi:10.1177/0022022107300276
9. Graesser, A.C., D'Mello, S.K., Strain, A.C.: Emotions in advanced learning technologies. In: Pekrun, R., Linnenbrick-Garcia, L. (eds.) International Handbook of Emotions in Education, pp. 473–493. Taylor and Francis, New York (2014)
10. Gross, J.J., John, O.P.: Individual differences in two emotion regulation processes: Implications for affect, relationships, and well-being. Journal of Personality and Social Psychology **85**(2), 348–362 (2003). doi:10.1037/0022-3514.85.2.348

11. Kavussanu, M., Dewar, A.L., Boardley, I.D.: Achievement goals and emotions in athletes: The mediating role of challenge and threat appraisals. Motivation and Emotion **38**, 589–599 (2014). doi:10.1007/s11031-014-9409-2
12. Lazarus, R.S., Folkman, S.: Stress, Appraisal, and Coping. Springer, New York (1984)
13. Lichtenfeld, S., Pekrun, R., Stupnisky, R.H., Reiss, K., Murayama, K.: Measuring students' emotions in the early years: The Achievement Emotions Questionnaire-Elementary School (AEQ-ES). Learning and Individual Differences **22**, 190–201 (2012). doi:10.1016/j.lindif.2011.04.009
14. Molina, P., Sala, M.N., Zappulla, C., Bonfigliuoli, C., Cavioni, V., Zanetti, M.A., Baiocco, R., Laghi, F., Pallini, S., De Stasio, S., Raccanello, D., Cicchetti, D.: The Emotion Regulation Checklist - Italian Translation. Validation of parent and teacher versions. European Journal of Developmental Psychology **11**(5), 624–634 (2014). doi:10.1080/17405629.2014.898581
15. Muthén, L.K., Muthén, B.O.: Mplus user's guide, 5th edn. Muthén & Muthén, Los Angeles (1998–2007)
16. Pekrun, R.: The control-value theory of achievement emotions: Assumptions, corollaries, and implications for educational research and practice. Educational Psychology Review **18**(4), 315–341 (2006). doi:10.1007/s10648-006-9029-9
17. Pekrun, R., Goetz, T., Frenzel, A.C., Barchfeld, P., Perry, R.P.: Measuring emotions in students' learning and performance: The Achievement Emotions Questionnaire (AEQ). Contemporary Educational Psychology **36**(1), 36–48 (2011). doi:10.1016/j. cedpsych.2010.10.002
18. Pekrun, R., Perry, R.P.: Control-value theory of achievement emotions. In: Pekrun, R., Linnenbrick-Garcia, L. (eds.) International Handbook of Emotions in Education, pp. 120–141. Taylor and Francis, New York (2014)
19. Pekrun, R., Stephens, E.J.: Academic emotions. In: Harris, K.R., Graham, S., Urdan, T., et al. (eds.) APA Educational Psychology Handbook. Individual differences and cultural and contextual factors, vol. 2, pp. 3–31. American Psychological Association, Washington, DC (2012)
20. Raccanello, D.: Students' expectations about interviewees' and interviewers' achievement emotions in job selection interviews. Journal of Employment Counseling **52**(2), 50–64 (2015). doi:10.1002/joec.12004
21. Raccanello, D., Brondino, M., Pasini, M.: Achievement emotions in technology enhanced learning: Development and validation of self-report instruments in the Italian context. Interaction Design and Architecture(s) Journal - IxD&A **23**, 68–81 (2014)
22. Raccanello, D., Brondino, M., Pasini, M.: On-line assessment of pride and shame: Relationships with cognitive dimensions in university students. Advances in Intelligent and Soft Computing **374**, 17–24 (2015). doi:10.1007/978-3-319-19632-9_3
23. Raccanello, D., Brondino, M., Pasini, M.: Two neglected moral emotions in university settings: Some preliminary data on pride and shame. Journal of Beliefs & Values: Studies in Religion & Education **36**(2), 231–238 (2015). doi:10.1080/13617672. 2015.1031535
24. Toepoel, V., Lugtig, P.: Online surveys are mixed-device surveys. Issues associated with the use of different (mobile) devices in web surveys. Methods, Data, Analyses **9**(2), 155–162 (2015). doi:10.12758/mda.2015.009

Erratum to: Methodologies and Intelligent Systems for Technology Enhanced Learning

**Mauro Caporuscio, Fernando De la Prieta, Tania Di Mascio,
Rosella Gennari, Javier Gutiérrez Rodríguez, Ricardo Azambuja-Silveira
and Pierpaolo Vittorini**

Erratum to:
M. Caporuscio et al. (eds.),
Methodologies and Intelligent Systems
for Technology Enhanced Learning,
Advances in Intelligent Systems and Computing,
DOI: 10.1007/ 978-3-319-40165-2

In the original version, the volume editor's name, and affiliation address were missing in the copyright page.

Ricardo Azambuja-Silveira
Department of Computer Science and statistics,
Federal University of Santa Catarina
Campus Universitário Cx.P. 476, 88040-900 Florianópolis S.C. (Brazil)

The online version of the updated original book can be found under
DOI: 10.1007/978-3-319-40165-2

© Springer International Publishing Switzerland 2016
M. Caporuscio et al. (eds.), *mis4TEL*,
Advances in Intelligent Systems and Computing 478,
DOI: 10.1007/978-3-319-40165-2_20

Correction to: Co-Robot Therapy to Foster Social Skills in Special Need Learners: Three Pilot Studies

Lundy Lewis, Nancy Charron, Christina Clamp, and Michael Craig

Correction to:
**Chapter "Co-Robot Therapy to Foster Social Skills in Special
Need Learners: Three Pilot Studies" in: M. Caporuscio et al.
(Eds.):** *Methodologies and Intelligent Systems for
Technology Enhanced Learning*,
https://doi.org/10.1007/978-3-319-40165-2_14

The book has been inadvertently published with incorrect version of figure in chapter 14, the figure 1 was revised and updated with the correct version.

The updated original version of this chapter can be found at
https://doi.org/10.1007/978-3-319-40165-2_14

Author Index

Printed in the United States
by Baker & Taylor Publisher Services